Elementary Particle Physics

....

Revealing the Secrets of Energy and Matter

Committee on Elementary-Particle Physics
Board on Physics and Astronomy
Commission on Physical Sciences, Mathematics, and Applications
National Research Council

NATIONAL ACADEMY PRESS
Washington, D.C.   1998

This project was supported by the Department of Energy under Grant No. DE-FG02-96ER40974, the National Science Foundation under Grant No. PHY-9600688, and a grant from the National Research Council's Basic Science Fund.

International Standard Book No. 0-309-06037-0
Library of Congress Catalog Card No. 97-81203

Additional copies are available from:

National Academy Press
2101 Constitution Ave., NW
Box 285
Washington, DC 20055
800-624-6242
202-334-3313 (in the Washington metropolitan area)
http://www.nap.edu

Front cover: The power of micro-vertex detectors, a new technology, is used to depict an interesting high-energy event. The detectors (shown in grey) are made of silicon and surrounded the collision point where high-energy interactions took place at 300,000 per second. The inner detector was located 3 cm from the collision point, and all detectors had spatial resolutions of about a thousandth of a centimeter. This fine resolution, needed to resolve the high density of tracks (some of which are shown in green and red), allows accurate extrapolation into the interaction region, inside the beam pipe—shown by the inner circle. The green tracks come from the original interaction, whereas the red ones come from two disconnected points. The latter are actually from B mesons that were created at the collision point but traveled several millimeters before decaying. The detector technology clearly reveals such decays even though the mean life is only a billionth of a second. From other information collected, one knows that this event is an example of the production of a pair of the very heavy top quarks, recently discovered by the CDF and D0 collaborations at Fermilab. (Courtesy of Joseph Incandela, CDF and Fermilab.)

Printed in the United States of America

The National Academy of Sciences is a private, nonprofit, self-perpetuating society of distinguished scholars engaged in scientific and engineering research, dedicated to the furtherance of science and technology and to their use for the general welfare. Upon the authority of the charter granted to it by Congress in 1863, the Academy has a mandate that requires it to advise the federal government on scientific and technical matters. Dr. Bruce Alberts is president of the National Academy of Sciences.

The National Academy of Engineering was established in 1964, under the charter of the National Academy of Sciences, as a parallel organization of outstanding engineers. It is autonomous in its administration and in the selection of its members, sharing with the National Academy of Sciences the responsibility for advising the federal government. The National Academy of Engineering also sponsors engineering programs aimed at meeting national needs, encourages education and research, and recognizes the superior achievements of engineers. Dr. William A. Wulf is president of the National Academy of Engineering.

The Institute of Medicine was established in 1970 by the National Academy of Sciences to secure the services of eminent members of appropriate professions in the examination of policy matters pertaining to the health of the public. The Institute acts under the responsibility given to the National Academy of Sciences by its congressional charter to be an adviser to the federal government and, upon its own initiative, to identify issues of medical care, research, and education. Dr. Kenneth I. Shine is president of the Institute of Medicine.

The National Research Council was established by the National Academy of Sciences in 1916 to associate the broad community of science and technology with the Academy's purposes of furthering knowledge and of advising the federal government. Functioning in accordance with general policies determined by the Academy, the Council has become the principal operating agency of both the National Academy of Sciences and the National Academy of Engineering in providing services to the government, the public, and the scientific and engineering communities. The Council is administered jointly by both Academies and the Institute of Medicine. Dr. Bruce Alberts and Dr. William A. Wulf are chairman and vice chairman, respectively, of the National Research Council.

Preface

The Committee on Elementary-Particle Physics (CEPP) was established by the Board on Physics and Astronomy as part of its decadal survey series *Physics in a New Era*. CEPP met six times over the course of 18 months, and it heard from program managers at the Department of Energy and the National Science Foundation and from congressional staff. The committee solicited input from the elementary-particle physics community through an email address, an Internet Web page, and a meeting after the 1996 Snowmass meeting of the American Physical Society's Divisions of Particles and Fields and Physics of Beams. At the seminar run by the International Committee for Future Accelerators at its October 1996 meeting in Japan, members of the committee initiated and participated in discussions on international collaboration.

CEPP was charged to describe what has been learned over the last two decades and to identify key physics objectives for the coming decades. The committee considered the facilities, instruments, and detectors that are required to carry out research in this field and outlined future options under realistic scenarios. The committee also outlined the field's relationships with other areas of physics and technology, and considered the general issues of education, manpower, and international cooperation; elementary-particle physics's relevance to society; its contributions to the welfare of the country; and the practical benefits of accelerator science and technology.

The committee would like to thank Donald C. Shapero and Robert L. Riemer from the Board on Physics and Astronomy for their efforts throughout the course of this study, attempting to steer its work toward a completed manuscript with the proper message, properly written. Katharine Metropolis edited parts of the report, and it is much to the better due to her efforts. The committee gratefully acknowledges the contributions of the following individuals who provided either material or particular advice that influenced its study: Jonathan A. Bagger, R. Michael Barnett, David G. Cassel, Gordon Cates, Ernie Fontes, Gerald Gabrielse, Christopher T. Hill, Joseph Robert Incandela, Judy Jackson, Andreas S. Kronfeld, Paul Langacker, Peter J. Limon, Yorikiyo Nagashima, Rene A. Ong, Michael Peskin, Nir Polonsky, Chris Quigg, Frank Sciulli, Stephen H. Shenker, Michael S. Turner, and Bill Willis. The committee also thanks Stephanie Selice, who edited the final production draft of the report.

The committee's work was supported by grants from the National Research Council's Basic Science Fund, the U.S. Department of Energy's Office of Energy Research, and the National Science Foundation's Physics Division. The committee thanks them for their support.

Finally, the committee would like to acknowledge David N. Schramm, who inspired this new survey of all fields of physics and was chair of the Board on Physics and Astronomy for most of the period of this study. We share with the physics and astronomy communities a deep regret for his untimely passing (shortly before this report was completed) and for the loss of his leadership.

Bruce Winstein
Chair
Committee on Elementary-Particle Physics

Acknowledgments

This report has been reviewed by individuals chosen for their diverse perspectives and technical expertise, in accordance with procedures approved by the National Research Council's (NRC's) Report Review Committee. The purpose of this independent review is to provide candid and critical comments that will assist the authors and the NRC in making the published report as sound as possible and to ensure that the report meets institutional standards for objectivity, evidence, and responsiveness to the study charge. The content of the review comments and draft manuscript remain confidential to protect the integrity of the deliberative process. We wish to thank the following individuals for their participation in the review of this report:

Robert K. Adair, Yale University
Lawrence M. Krauss, Case Western Reserve University
Leon Lederman, Fermilab
Francis Low, Massachusetts Institute of Technology
Michael Riordan, Stanford Linear Accelerator Center
John Schwarz, California Institute of Technology
Sam B. Treiman, Princeton University
Edward Witten, Institute for Advanced Study, Princeton University

Although the individuals listed above have provided many constructive comments and suggestions, responsibility for the final content of this report rests solely with the authoring committee and the NRC.

Contents

Executive Summary   1

1 *Introduction*   16

2 *What Is Elementary-Particle Physics?*   21
Introduction, 21
Fundamental Constituents of Matter, 22
    Earth, Air, Fire, and Water, 22
    Chemical Elements, the Periodic Table, and Atoms, 22
    Protons, Neutrons, and the Electron, 23
    Today's Fundamental Constituents, 23
    The Neutrino and Leptons, 23
    Particle "Generations," 24
Relativity, Quantum Mechanics, and Particle Accelerators, 25
Forces, 26
    Gravity, 27
    Electric and Magnetic Forces; Electromagnetism, 27
    Weak and Strong Forces, 28
    What "Transmits" Forces?, 29
    Unification of Forces, 30
Laws of Nature, 30
Particle Collisions, 31
    Scattering Experiments, 31
    Colliders, 31
Summary, 32

3   *Symmetries, Forces, and Particles*   33
        Introduction, 33
        Symmetries, 34
            Symmetries and Particle Physics, 35
            Local or Gauge Symmetries, 37
        The Standard Model, 39
            Spontaneous Symmetry Breaking, 42
            Higgs Boson, 43
            Generation-Changing Interactions, 43
        Beyond the Standard Model, 45
            Symmetry Breaking and Supersymmetry, 45
            Grand Unification, 47
            Why Are There Three Generations?, 48
        Physics of the Planck Scale, 49
            String Theory, 50

4   *The Past 25 Years: Establishing the Standard Model*   52
        Introduction, 52
        The World of Elementary-Particle Physics Circa 1972, 52
        The Forces, 53
            The Electroweak Force, 53
            The Strong Force, 56
        Constituent Particles, 59
            Discovery of the Charm Quark, 59
            Discovery of the Tau Lepton, 59
            Discovery of the Bottom Quark, 61
            Discovery of the Top Quark, 62
            Counting the Number of Generations, 63
        Particle-Antiparticle Asymmetry, 63
        Other Studies, 65
            Measuring the Mass of Neutrinos, 65
            Searching for Proton Decay, 66
            Other Physics Beyond the Standard Model, 66
        Summary, 66

5   *The Physics of the Next Decade*   68
        Overview, 68
        What Is the Origin of Mass?, 70
        Why Are There Energy Scales that Are So Vastly Different?, 71
        What Is the Origin of Matter-Antimatter Asymmetry?, 72
        Patterns of Quark and Lepton Masses and Transitions, 73
        Understanding the Strong Force, 75
        Are There Unexpected Phenomena?, 76
        Summary, 77

6   *Accelerators and Detectors: The Tools of Elementary-Particle Physics* 78
       Introduction, 78
       Particle Accelerators, 84
           Performance of Existing Accelerators, 84
           Accelerator Facilities Under Construction, 86
           Options for Future Facilities, 88
       Detectors in Elementary-Particle Physics, 91
           Particle Detection, 92
           Particle Detector Topologies, 93
           Challenges for the Next 10 to 20 Years, 98

7   *The Role of New Facilities*   101
       Overview, 101
       The Landscape in 2010, 102
       Future Colliders, 104
           The Physics Need, 105
           Colliders to Address the Physics Need, 105
       The Next Steps, 108

8   *Accelerator-Detector Technology and Benefits to Society*   110
       Introduction, 110
       The Machine Frontier, 111
       Synchrotron Radiation: Using X-Ray Light to See the
           World in Atomic Detail, 111
       Science and Industry in a Partnership Down to the Wire, 114
       The Detector Frontier, 115
       The Computing Frontier and Elementary-Particle Physics, 116
       Technologies for the Next 20 Years, 119

9   *Interactions with and Connections to Other Branches of Physics
       and Technology*   121
       Introduction, 121
       Cosmology, 121
           Dark Matter, 122
           Structure Formation, 125
           Baryogenesis and Nucleosynthesis, 126
       Astrophysics, 127
           Physics of the Sun, 127
           Supernovas, 127
           Cosmic Rays, 127
       Nuclear Physics, 129
       Atomic Physics, 130
       Condensed-Matter Physics, 131

Fluid Dynamics, 133
Mathematical and Computational Physics, 133

10 *Elementary-Particle Physics in Today's Society*   135
Introduction, 135
Historical Background, 135
    Particle Physics Until World War II (the First 50 Years), 135
    Particle Physics After World War II (the Second 50 Years), 136
    Impact of the Termination of the Superconducting Super Collider, 137
Organizational Structures, 139
    Universities, 140
    Laboratories, 140
    Experimental Collaborations, 141
    The Advisory System, 142
    International Cooperation, 143
    Future Challenges, 144
Education in Elementary-Particle Physics, 148
    Particle Physics Graduate Education, 148
    Outreach to the Public, 149

11 *Conclusions and Recommendations*   151
Introduction, 151
Recommendations for U.S. Elementary-Particle Physics, 153
    1. Recommendations Concerning the High-Energy Frontier, 154
    2. Recommendation for Addressing Important Fundamental
        Physics Problems Below the TeV Mass Scale, 158
Conclusion, 159

Appendix *Glossary, Abbreviations, and Acronyms*   161

Elementary-Particle Physics

Executive Summary

INTRODUCTION

The Committee on Elementary-Particle Physics was convened by the National Research Council's Board on Physics and Astronomy to assess the field of elementary-particle physics as part of the survey series *Physics in a New Era*. The committee was charged to make recommendations about the role that the United States should play in research in this field in the next two decades. The members of the committee, all active researchers in the field, brought a diversity of perspectives to bear on this study.

In preparing this report, the committee had two main objectives: (1) to describe the current status of elementary-particle physics and the most important research issues within this domain; (2) to identify the elements of a research program for the next two decades that, given limited resources, represents a wise approach to addressing these issues and maintaining the United States as a leader in the field.

ELEMENTARY-PARTICLE PHYSICS: EXPLAINING THE PHYSICAL WORLD

How it is that our universe came to be so rich and varied? Why are there stars, light, planets, and a hundred different atoms that can be combined into countless molecules? Elementary-particle physicists seek answers to these questions by studying subatomic particles and forces. Although these investigations require sophisticated instruments to reveal phenomena far smaller and more en-

1

ergetic than we are aware of in daily life, the deep connection between the two realms inspires researchers in elementary-particle physics and lends added significance to their investigations. In fact, the properties and interactions of the elementary particles have much to say about the properties of the world around us.

A century ago, the first elementary particle—the electron—was identified. A revolutionary view of the way matter in the universe is put together was provided by experimental evidence that electrons were basic constituents of all atoms and that they carried electricity. The theory of quantum mechanics explained the paradoxical motion of electrons in the atom and the formation of molecules. Eventually, a vast range of phenomena—the stiffness of steel, the way gasoline burns, the colors in the surface of a bubble, the ability of x rays to reveal tumors inside the body—could be accounted for by the quantum mechanical behavior of the electron. This new perspective, a view of the world on a more fundamental scale than the everyday world of our experience, led to a century of spectacular science and technological innovation. Understanding the behavior of the electron and the photon (the quantum of light) has been critically important to the fields of chemistry, materials science, and biology, as well as to the development of modern computing and communication.

Particle physicists further zoomed in on the subatomic realm with increasingly powerful instruments. Forces were revealed on the subatomic level that no one had predicted, the best example being the strong nuclear force that holds the atomic nucleus together. Experiments revealed the existence of hundreds of different—and unexpected—particles. Eventually patterns emerged and theories were put together and tested; today, elementary-particle physics provides the basis for understanding an astonishing variety of phenomena—including those in our daily lives—in terms of just a few truly elementary particles and the forces between them.

The remarkable state of our understanding of elementary particles, embodied in the present theory called the "Standard Model," has taken shape over the last 30 years. The Standard Model provides an organizing framework for the known elementary particles. These consist of "matter particles," which are grouped into "families," and "force particles." The first family includes the electron, two kinds of quarks (called "up" and "down"), and a neutrino, a particle released when atomic nuclei undergo radioactive decay. There are two more families consisting of progressively heavier pairs of quarks and a corresponding lepton and neutrino. All normal, tangible matter is made up only of particles from the first family, since the others live for very short times. Why are there three families? This question is one of the great unsolved mysteries of elementary-particle physics.

The matter particles exert forces on one another that are understood as resulting from the exchange of the force-carrying particles. Electric and magnetic forces arise when particles exchange photons (the familiar repulsion or attraction of two magnets results from one of them emitting photons that the other receives). The strong force that holds quarks together to form protons and neutrons comes from the exchange of gluons. The weak forces that cause radioactive decay are created by massive *W* and *Z* particles (the photon and gluon have no mass). These three forces have been successfully described by quantum theories that have remarkably similar structures.

Yet crucial questions remain unanswered by the Standard Model. For example, the masses of elementary particles can be established by measurement, but they appear to be arbitrary. There is not a set of rules that allows these masses to be calculated or that explains why the up quark is a bit lighter than the down quark. However, the consequences of these values are profound. Were it the other way around, creation of the heavier elements in the interior of stars—elements essential for the existence of planets and life—would have been dramatically different, leading to a far duller universe.

ELEMENTARY-PARTICLE PHYSICS AND SOCIETY

Elementary-particle physics is basic research, driven by intellectual excitement and the desire to understand the underlying structure of the universe. Its discoveries illuminate all of science, and the technology developed in the course of this research may ultimately be applied for practical benefit.

• Synchrotrons were developed to accelerate particles, cause collisions that create new particles, and provide clues about their interactions. A by-product of accelerating particles is the production of intense electromagnetic radiation from the visible part of the spectrum all the way to x rays. Several laboratories now operate synchrotrons purely for the purpose of generating such radiation; they are invaluable for researchers in surface chemistry, materials science and engineering, environmental science, and biology. Biological applications are growing at a rapidly accelerating pace and promise to give new insights into living systems.

• Devices and techniques developed for elementary-particle physics research are important in several medical imaging techniques. Computer-aided tomography (the CT scan) and positron-emission tomography (the PET scan) use detectors largely developed for particle physics experimentation. Development of the industrial capability to produce large quantities of high-quality superconducting wire, in order to meet the demands of particle accelerators, led directly to the billion-dollar world market in this wire, primarily for use in magnetic resonance imaging (MRI).

• The World Wide Web, which was developed to enable elementary-particle physicists around the world to share information quickly and easily, now gives every school with a computer access to the largest library of information on the globe.

These and other offshoots have been immensely valuable and have had a profound impact on other sciences and on our society. Elementary-particle physicists take pleasure in making these contributions for the good of society, but their main goal is to understand the universe: why it looks the way it does today, how it evolved from the earliest moments, and what its ultimate fate will be. The intellectual significance of the field is reflected in the number of Nobel Prizes awarded to elementary-particle physicists, in the illumination that elementary-particle physics has provided to other branches of science, and most important, in the new picture it is developing of the way in which fundamental particles and forces shape our world.

National support of endeavors such as astronomy and elementary-particle physics is dedicated to the proposition that deepening our knowledge of the world we inhabit increases the pleasure, richness, and value of life. When a nation takes pride in contributing to such explorations, it says something important about itself.

KEY RESEARCH OPPORTUNITIES

Recent advances in technology, experimental techniques, and theoretical understanding mean that over the next two decades, it will be possible to investigate some of the most compelling issues in particle physics. Questions that once seemed beyond the scope of science now lie within the reach of experiment, and it will be possible to achieve a deeper understanding, not only of elementary particles, but also of the earliest moments in the history of the universe.

Determining What Gives Elementary Particles Their Mass

One especially important opportunity is to understand what determines the observed—and very disparate—masses of both the force particles and the matter particles. The Standard Model of particle physics invokes a very special kind of particle, called the Higgs (after the theorist who suggested it). Higgs particles (there may be one or more kinds) are unlike any other particles. Their effects are ever present, even in the vacuum; they give all matter particles their mass and account for the large mass of the carriers of the weak force and the masslessness of the carrier of the closely related electromagnetic force. (Physicists refer to this important asymmetry as electroweak symmetry breaking.) Other possible explanations involve new, very strong forces.

Only experiments at higher energies than those to which experimenters now have access will conclusively determine which, if any, of these theories is actually realized in nature. One thing is certain: The creation of mass involves some completely new physics that must show up in experiments at sufficiently high energy. There are very compelling theoretical arguments guaranteeing that this physics will be uncovered by the next generation of experiments.

Supersymmetry and Strings

A long-standing goal of science has been to find the simplicity underlying the wide diversity in nature. Two examples of such achievements in basic physics were the realization that sound and heat could be understood in terms of the motion of atoms, and Maxwell's synthesis of electricity, magnetism, and light, in terms of the electromagnetic field. These unifications of seemingly disparate phenomena resulted in a far deeper, far more useful understanding.

Theories have recently been developed that link elementary particles, the very smallest known structures, with one of the grandest questions of all: How did the universe begin? These theories suggest that many of the complexities manifest at lower energies would be greatly simplified under conditions of extremely high energy. For example, under such conditions the different forces between the particles would be seen as unified, different manifestations of a single underlying force. There is nothing artificial about these energies: They are understood to have prevailed at the time of the "big bang" that initiated the expansion of the universe. If these theories are correct, the simplicity long sought by particle physicists is a fact of cosmic history.

Although re-creating these energies lies beyond the reach of existing and planned accelerators, experiments at these accelerators could nonetheless reveal evidence for what is called supersymmetry. Supersymmetry, which is really a very profound statement about the structure of space and time, predicts that for every fundamental particle there should exist another related and as yet undiscovered new particle; conclusive evidence for these additional particles is eagerly sought because supersymmetry shows how the electroweak and strong interactions could be unified.

In addition, evidence of supersymmetry would also support another even more comprehensive theory, called string theory. Traditionally, elementary particles have been modeled as points that take up no space at all. This approach leads to some theoretical problems because two particles could (in principle) get extremely close and exert arbitrarily large forces on each other. String theory solves this problem by picturing particles as extremely tiny vibrating loops, with the details of their vibrations determining their properties and interactions. This simple idea, with the aid of recent theoretical developments, leads to a theory that is able to encompass *all* of the forces of nature in a unified and self-consistent manner, including—for the first time—gravity.

Role of Antimatter in the Universe

Deep mysteries are connected with the way particles from different families transform into one another. For example, a tiny but startling asymmetry in the behavior of matter and antimatter has been found in certain particle interactions. These particle decays violate what is called CP or time-reversal symmetry. This curious phenomenon is crucial to knowing how matter came to predominate in the universe because without it, there would be no stable chunks of matter in the form of stars, planets, and ultimately, human beings to wonder at it all. However, there is as yet no fundamental understanding of this important asymmetry. Experiments in the coming decade are poised to increase greatly our knowledge in this area.

Of course, new understanding does not always proceed along a direct path from prediction to discovery. The history of particle physics is full of unexpected experimental results, which have lighted the way to more useful and complete models. Historically, such important surprises have been most probable when experiments are conducted at previously inaccessible energies.

ACCELERATORS:
INTERNATIONAL INSTRUMENTS OF RESEARCH

The Role of Accelerators

The scale of elementary particles is so far removed from the human scale that it is almost impossible to comprehend. Elementary-particle physicists today are exploring phenomena on a scale as small as 10^{-18} m. An object this size compares in length to a meter stick as a meter stick compares to the distance light travels in 100 years. It is amazing that there is a way to study objects this small, but it can be done with particle accelerators, the "microscopes" of particle physics. Accelerators have become so complex and expensive that national, and increasingly international, collaborations are required to design, build, and operate them.

When particles of sufficiently high energy collide, new particles are created out of the energy of the collision. The higher the energy of the collision, the more massive are the particles it can produce. There are strong theoretical arguments that the key to understanding some of the most important issues before elementary-particle physics today is attaining a high rate of collisions in the tera-electron-volt (TeV; 10^{12} eV) range, today's energy frontier.

Research and Development

Over the past half century, particle accelerators, particularly machines that bring two beams of particles into head-on collision (colliders), have been increasingly important to elementary-particle physics. Without them, it would have been impossible to obtain most of the data that led to the development of the Standard Model, and the important and beautiful structures of matter at layers below the visible world would have remained hidden. The energy of accelerators has been increased spectacularly over this period, not simply by enlarging or improving a given design but as a result of many innovations and technological breakthroughs by engineers and accelerator physicists. Where the exact limits of the current types of machines lie is debated, but it is clear that to move substantially beyond the capabilities of present technologies in a cost-effective manner, we must continue to find new ways to accelerate particles. Accelerator development has created a sophisticated technology with uses in other areas of science, chief among which is the synchrotron light source mentioned earlier, and continued development can reasonably be expected to make broader contributions of this kind.

THE CURRENT U.S. PROGRAM

The scientific strength of the United States in the field of elementary-particle physics is manifest in the quality and influence of the research it carries out. Members of this community, traditionally some 2,000 strong, have played important and leading roles in obtaining incisive experimental results, coaxing innovative technologies into existence, and developing important breakthroughs in theory. One measure of excellence is the fact that many of the best students in the world choose to come to the United States for their graduate training. Another is the leading role that U.S. physicists currently play in preparing and executing experiments aimed at addressing many of the most significant research questions in the field.

Research at Accelerator Laboratories

The present generation of facilities includes the Stanford Linear Accelerator (SLAC) and the Fermi National Accelerator Laboratory (FNAL). SLAC is a 2-mile-long linear electron accelerator that is capable of accelerating electrons to an energy of 50 GeV (10^9 eV), one million times more energy than acquired by electrons in the picture tube of a television set as they travel from the electron gun to the screen. At FNAL, protons are accelerated in the 4-mile circular Tevatron to an energy of 1,000 GeV. This is the most powerful accelerator now in operation, using approximately 1,000 superconducting magnets to steer particles along the desired paths.

The United States is also home to the Cornell Electron Storage Ring (CESR), the highest-luminosity collider ever built, an electron-positron collider operating at approximately 5 GeV per beam, and the Alternating Gradient Synchrotron (AGS) at Brookhaven National Laboratory, a 30-GeV proton accelerator with the highest beam intensity in the world available for elementary-particle physics.

U.S. researchers are currently conducting experiments at major overseas accelerators, including the Large Electron-Positron collider (LEP) at CERN (the European Laboratory for Particle Physics) near Geneva, the highest-energy electron-positron collider ever constructed; HERA, the world's only electron-proton collider, at the DESY laboratory in Hamburg; and modest-energy electron colliders in Beijing and Novosibirsk. They will resume working at Japan's KEK laboratory in the very near future.

An important aspect of experiments at all these accelerators is the great complexity and sophistication of the apparatus at the business end of the accelerator. Very high energy collisions generate a vast profusion of particles. To separate out the interesting events requires complex systems of detectors to trace the paths of the particles, using extremely high-speed electronics to evaluate the events in real time. All of this equipment must have capabilities that far exceed those available commercially. The processing power of the custom high-speed electronics used to untangle the massive bursts of data that cascade out of the detectors compares with the capabilities of the fastest supercomputers.

Research with Particles from Space

Accelerator-based research is complemented by important studies that detect particles produced in space. One such study addresses the flux of neutrinos emitted from the interior of our Sun as it burns. Over the last two decades, careful measurements with enormous detectors at several independent sites have revealed that the number of neutrinos detected is smaller than the number predicted. Many of the speculations that attempt to explain the discrepancy have significant implications for our picture of elementary-particle physics.

A question that has long troubled astrophysicists is the composition of 90% of the mass of the universe, which is invisible—its presence betrayed only by its gravitational effect on the behavior of galaxies. One exciting possibility is that this so-called dark matter, the focus of a second group of studies, consists partly of particles predicted by the theory of supersymmetry.

The third study is concerned with the origin of the highest-energy cosmic rays that must originate in cosmic events of extreme violence. These could be produced by some unknown fundamental process, possibly involving relics of the early universe. When they strike Earth's atmosphere, they generate a whole cascade of particles that show up in large arrays of detectors on Earth's surface.

Over the next two decades, questions of the greatest importance to understanding the universe at its most fundamental level will at last come within reach of experiment. How elementary particles acquire their mass and whether the known forces are simply manifestations of a single underlying force are two of the most significant issues that must be addressed in comprehending the world around us. It is deep and profound questions such as these that first capture the imaginations of bright young people, whether or not they work in particle physics, leading them to and sustaining them on the challenging and difficult road to a technological education. These are the very bright young people that eventually become our scientists and engineers.

The committee believes that these issues are sufficiently compelling that the

U.S. particle physics community should play a leading role in the international endeavor to conduct research capable of addressing them. If the recommendations in this report are adopted, the United States can be at the forefront of this profound and fascinating intellectual adventure.

RECOMMENDATIONS FOR
U.S. ELEMENTARY-PARTICLE PHYSICS

The committee has developed its recommendations with two goals: (1) to exploit the great opportunities for discovery that lie ahead and (2) to maintain U.S. leadership in the field of elementary-particle physics. These goals require a diverse but focused program.

We are poised on the threshold of a new energy frontier, where discoveries are certain to be made and new phenomena are likely to be revealed. This is the TeV mass scale, where both well-established theory and revolutionary ideas predict new physics. First, the remarkable success of the Standard Model ensures that the secret of electroweak symmetry breaking will be revealed at this scale. Second, the exciting idea of supersymmetry, which offers the hope of great insights into unification of all the forces of nature, predicts that a rich array of new particles can be produced. Finally, we will obtain the first glimpse of physics well above the typical mass scale of the Standard Model. In the past, when such a large step has been taken, dramatic experimental surprises have occurred. One might expect that similar revolutionary discoveries will be made at the TeV mass scale.

The committee therefore believes that the highest priority is full involvement in TeV mass scale physics at large facilities uniquely suited to this purpose. This involvement includes exploiting the Fermilab Collider (presently the highest-energy facility extant); strong participation in construction of and research at the Large Hadron Collider (LHC) being built in Europe; and taking a leadership role in a future forefront international facility, possibly to be built in the United States. This path has historically provided the most fruitful avenue for uncovering new phenomena.

Other problems of great importance to the understanding of elementary particles do not require the highest energies for elucidation. One is understanding rare quark and lepton transitions. Another is the nature of CP violation—a phenomenon that bears on the apparent dominance of matter over antimatter in the universe. There are additional astrophysical questions of great importance that can likely be explained by particle physics dynamics, the most important being the nature of dark matter. A number of the most important findings in the field in the past two decades have been made by experiments studying problems such as these, and facilities presently being upgraded or under construction will allow such studies to continue. The committee believes it is crucial to support a well-targeted program in these areas. Given the limited resources that will be available, however, maintaining a proper balance between such efforts and those at the energy frontier will require difficult choices and keen foresight.

The committee's recommendations are therefore grouped into two classes: first, those relevant to the energy frontier, and second, one concerning important studies that are best done elsewhere. Both are essential to a balanced program.

Before presenting its recommendations concerning experimental initiatives, the committee comments on two subdisciplines of the field that are critical elements of a forefront program: non-facility-specific advanced accelerator R&D, which can lead to extension of the energy frontier, and theoretical physics, which provides the framework that organizes our observations.

Advances in elementary-particle physics have historically been tied to advances in accelerator technology. Accelerator research and development is of two general types—efforts targeted at the design and construction of specific facilities and more generic (and forward-looking) R&D targeting completely new methods of acceleration that will be required to support energy frontier facilities decades from now, should the physics demand it. This report contains specific recommendations with regard to the former. It is necessary to maintain an appropriate level of investigation in the latter area to secure the longer-term future of the field.

Theoretical work in elementary-particle physics provides the intellectual foundation that motivates and interconnects much of experimental research. The more formal areas of theoretical physics, especially string theory, hold the promise of providing a picture of the universe that accounts for an extremely broad range of observations and phenomena. The committee believes that a healthy level of activity both in formal areas and in the more phenomenological investigations that touch directly on experiments now and in the coming decade should be maintained.

1. Recommendations Concerning the High-Energy Frontier

At the present time, the Tevatron at Fermilab and the Large Electron-Positron collider (LEP II) in Geneva are the only machines operating at the energy frontier. In two years, LEP II will be dismantled, leaving the Tevatron alone at this frontier until completion of the LHC in the middle of the next decade. The LHC will dramatically extend the energy reach, pushing beyond the TeV scale, where we know that the physics of electroweak symmetry breaking must appear. However, this report concludes that in the future, another collider will be required to complement or extend the range of the LHC and to explore fully the physics of the TeV scale. These considerations motivate a chronological structure for the committee's recommendations concerning the high-energy frontier.

1.a. Recommendation on the Fermilab Collider Facility

The United States should capitalize on the potential of the Fermilab Collider Facility while it has unique capabilities for investigations of high mass scale physics.

The Tevatron collider is the highest-energy accelerator in the world today and will remain so until the LHC era. The recent discovery of the top quark at this facility demonstrates its power to explore physics that is otherwise inaccessible. Its capabilities will be considerably enhanced with the new Main Injector. Although the LHC will be the first machine to extensively explore electroweak symmetry breaking, some of the new particles associated with the TeV scale might exist within the reach of the Tevatron. In particular, the upgraded Tevatron collider facility might discover supersymmetry. This would dramatically enhance our understanding of the universe.

1.b. Recommendation on the Large Hadron Collider

The committee enthusiastically endorses U.S. participation in the Large Hadron Collider project as a vital and essential component of the U.S. experimental particle physics program.

In the middle of the next decade, the LHC will supersede the Tevatron Collider as the highest-energy machine in the world. U.S. physicists, with their extensive experience at Fermilab and in the research and development toward construction and use of the Superconducting Super Collider (SSC), have established critical roles in the construction of the LHC machine and of the two largest experiments. The resources involved have been established in an agreement reached in 1997 by the Department of Energy, the National Science Foundation, and CERN, the host laboratory.

The LHC will systematically explore a new energy regime, the TeV mass scale. LHC experiments will elucidate the mechanism of electroweak symmetry breaking, the central question of elementary-particle physics. The experiments will decisively test the prediction that a rich array of supersymmetric particles appears at this mass scale. If supersymmetry is indeed present at the TeV scale, the LHC will initiate the exploration of a vast new world.

The committee is convinced that participation in the enormously exciting physics promised by the LHC is essential for the vitality and continuity of U.S. particle physics. The committee also believes that U.S. participation is vital for the success of the project.

1.c. Recommendations on the Next Generation of Accelerators

As this report emphasizes, the committee anticipates that major discoveries will be made at the LHC. These will almost certainly point toward new phenomena that physicists will want to explore using an appropriate new collider.

Three types of machines have been discussed by the physics community: electron-positron linear colliders, muon colliders, and very large hadron colliders. Each has its unique capabilities and challenges, and each is at a different stage of development. Only the linear collider is far enough along to proceed to a conceptual design, with the engineering details and cost and schedule information appropriate to this stage. The other two options are sufficiently promising that increased research efforts are called for to make more realistic preliminary designs. These steps will put the community in the position to make a decision in the future about starting a new collider construction project with the best information possible.

A collider that complements or extends the reach of the LHC will require multiyear and multinational cooperation because of the magnitude of the resources needed. If the United States is to maintain a leadership role in this enterprise, it must participate both in accelerator technology development and in international decisions on the choice of technology and the location of the next facility. Although it is highly desirable to have a forefront facility located within the United States, it is crucial that the United States maintain a technological base sufficient to allow full participation in all aspects of the design, construction, and operation of such a facility, independent of its ultimate location.

Recognizing that it is too soon to endorse construction of any new machines, the committee makes recommendations concerning the development of each.

1.c.1. Recommendation on Electron Colliders

The committee recommends support of an international effort leading toward a complete design and cost estimate of an electron linear collider that would be able ultimately to reach a center-of-mass energy of 1.5 TeV and a luminosity of 10^{34} cm^{-2} s^{-1}.

An electron linear collider would contribute important measurements complementary to those from the LHC toward understanding the fundamental physics of the TeV mass scale. In the past, lepton colliders have been essential complements to hadron colliders. For example, W and Z bosons were discovered in a hadron collider, but many of their important properties could be determined only with the electron-positron colliders at LEP and the Stanford Linear Collider (SLC). For the physics of the TeV scale, this complementarity will likely continue to be important.

Laboratories in the United States, Japan, and Europe have been engaged for many years in research and development on an electron linear collider operating with an energy of 1 TeV or more. Stanford Linear Accelerator Center (SLAC), with its unique expertise in linear collider technology and the experience gained through the construction and operation of the SLC, is playing a critical leading role in these efforts. Many of the systems required for a second-generation linear collider have been or are being demonstrated. The next natural step is a complete design report, accompanied by cost optimization studies and a complete cost estimate.

The committee encourages the U.S. linear collider community to work co-operatively with international partners on the development of a common design and possible management structures.

The effort to complete an electron linear collider design and optimized cost estimate could be finished early in the next decade. It will then be necessary for the United States, together with the international elementary-particle physics community, to consider a number of factors in deciding whether to propose construction:

- The physics case for such a collider in light of any new discoveries in the intervening years;
- The construction and operating costs of the facility, together with the commitments and plans of the nations interested in hosting or participating in a linear collider; and
- The status of development of muon and hadron colliders.

1.c.2. Recommendation on Muon and Hadron Colliders

R&D targeted at developing the technologies for muon and very large hadron colliders should be vigorously pursued.

Experiments at the LHC may indicate new physics at energy scales significantly beyond those that it can decisively reach. Extension of the energy frontier beyond the reach afforded by the LHC will require the development of new technologies. A muon collider or a very large hadron collider has the potential for supporting even higher energies and luminosities in the post-LHC era. R&D efforts in both of these areas are in the early stages. Muon collider technology remains to be demonstrated, so the need is to focus on the development and validation of concepts. Present-day hadron collider technology could likely be used to construct a facility with a reach significantly beyond LHC, but the cost would be prohibitive. Hence efforts in this realm should focus on a reduction of cost through the use of advanced technologies. Development of both muon and hadron collider technologies must be pursued in a timely fashion to determine whether they represent technologically and economically viable options for reaching energies beyond those explored with the LHC.

2. Recommendation for Addressing Important Fundamental Physics Problems Below the TeV Mass Scale

The committee recommends strong support for a well-targeted program to study the fundamental particle physics that can best be explored with experiments below the TeV scale.

In its first recommendation, the committee has emphasized the range of important physics questions that are addressed at the TeV scale. It is important to recognize, however, that a number of outstanding fundamental questions can best be studied using other techniques. Foremost among these are the understanding of quark and lepton flavor mixing and of particle-antiparticle asymmetry (CP violation). There are also astrophysical questions of importance to particle physics, such as the nature of dark matter.

Experiments studying rare transitions between different families of quarks or leptons are extremely sensitive to new and interesting physics. For example, the 1964 experiment discovering CP violation found new fundamental physics that we are still trying to understand. One of the major themes of experimental particle physics in the next decade will be a systematic study of the interactions that mix the families of quarks and leptons.

Experiments in this area include several categories:

• Decays of the bottom quark. The central question to answer is whether CP violation is explained within the framework of the Standard Model or whether it is due to some new physics. The Standard Model explanation makes specific predictions that can be tested with very large samples of B mesons.

• Decays of the strange quark. Although CP violation was discovered in the decays of K mesons containing the strange quark, there are still outstanding issues in the CP-violating decays of strange particles. Experiments using extremely intense kaon beams give unique information about CP violation.

• Neutrino oscillations. Many experiments now give hints that a neutrino of one family can change into one of another family. One of the most important discoveries possible in the next decade would be unambiguous confirmation of any one of these hints.

A new era of research in these areas will begin in the next few years as experiments that should decisively answer many of the long-standing questions come on-line. Key U.S. facilities—the Positron-Electron Project II (PEP-II) and the CESR upgrade in addition to the Main Injector—will begin operations in the next few years with greatly enhanced capabilities to address this very important physics.

It is important to operate the newly built facilities and fund their critical experiments at the level required to take advantage of the physics opportunities they present. Historically, the U.S. high-energy physics community has phased out programs to accommodate those that are more scientifically desirable, and it should continue to do so. Because of limitations in resources for the field worldwide, in the future, only initiatives that have the most promise for scientific advancement should be undertaken.

CONCLUSION

The recommendations above, if adopted, should maintain U.S. leadership in the field of elementary-particle physics well into the next century. They will allow our scientists to participate in what are likely to be profound and exciting discoveries, discoveries of a nature not seen before.

1

❖

Introduction

This report assesses the field of elementary-particle physics. The Committee on Elementary-Particle Physics of the National Research Council's Board on Physics and Astronomy was assembled to review what has been learned, to identify research priorities for the next two decades, and to describe the instruments and infrastructure needed to carry them out. This chapter introduces the main themes of the report.

The universe is constructed with remarkable economy. Galaxies and hummingbirds, computers and the neurons firing in our brains as we read this sentence—everything in the tangible world is built from about a hundred different kinds of atoms. Every atom, in turn, is a combination of just three different constituents: u quarks and d quarks (which in different combinations form protons and neutrons) and electrons. Up to the resolution of current experiments, no internal parts have been detected in quarks and leptons, so they are called elementary particles.

Although elementary particles are infinitesimal—smaller relative to a grain of sand than a grain of sand is to the entire Earth—the consequences of their properties are enormous. If, for example, the electron were much heavier, the universe would have evolved entirely differently: No atoms would exist, and the universe would now consist solely of electrically neutral particles. No stars would shine; no people would be around to wonder at the universe's origin or ultimate fate.

The richness of the phenomena in our universe, even biological systems, stems from the physical principles that operate on the scale of elementary particles. Investigating these particles is, in effect, deciphering the genetic code for the universe: why it is the way it is and how it came to be that way. The goal of elementary-particle physics is to understand the world around us by identifying the elementary particles, understanding their properties, and learning how they interact.

Researchers proceed toward this goal along two avenues: (1) by conducting experiments and (2) by trying to determine the physical principles that account for the phenomena they observe—what theoretical physicist Richard Feynman called "the patterns [in] the phenomena of nature [that are] not apparent to the eye, but only to the eye of analysis." The dialog between experimenters and theorists shapes the research priorities of the field: Experimental research is often guided by theoretical predictions; about as often, phenomena will turn up in experimental data that no one expected to find, and theorists endeavor to account for them.

Investigating phenomena on this almost unimaginably minute scale requires the most powerful microscopes ever built: devices known as particle accelerators. In a particle accelerator, beams of subatomic particles are boosted to nearly the speed of light and then brought into collision with either a stationary target or another beam of accelerated particles coming head-on. In these collisions, remarkably, matter is actually created. The particles that emerge from the collision point, like sparks radiating out from microscopic exploding fireworks, are not contained within the original colliding particles. They are created out of the energy of the collision according to the rules of relativistic quantum mechanics. The higher the energy of the collision, the heavier are the particles it can create. Such particles, although fundamental, are often ephemeral, existing only briefly before transforming themselves into more stable particles. High-energy accelerators thus provide elementary-particle physicists with the opportunity to study phenomena that they could otherwise not observe on Earth. Today's accelerators can collide particles with such high energies that, on a very small scale, they replicate the conditions prevailing when the universe was only a fraction of a second old and enable physicists to study the kinds of particles that long ago shaped the evolution of the universe, before the cosmos cooled off too much for these particles to continue to be produced.

If accelerators function as microscopes, then the eyes and brains that see and record the phenomena that accelerators reveal are detectors. In essence, detectors are devices that surround the collision point to capture enough information about the particles produced to deduce their properties: Are they electrically charged? Are they light or relatively massive? How long do they exist before being transformed into other kinds of particles?

Over the past hundred years, advances in experimental instrumentation and technique have revealed subatomic phenomena that scientists in earlier centuries had no idea existed. These phenomena, in turn, have led to discoveries of physical principles that are crucial for understanding how the universe is put together.

In addition, these increasingly sophisticated ventures in both experiment and theory have opened profitable new avenues in many fields. Last year, for example, marked the one-hundredth anniversary of the first discovery of an elementary particle. In 1897, British physicist J.J. Thomson concluded that in experiments with cathode-ray tubes, he had seen negatively charged constituents of atoms. Thomson called these entities "corpuscles"; other physicists referred to them by the name that stuck: "electrons." Around the same time, in an enormously fruitful period of research into the nature of matter and energy, x rays and radioactivity were also discovered. When classical concepts of physics proved incapable of explaining these phenomena, quantum mechanics was developed.

From this work emerged nothing less than a radically new picture of nature, which in turn had dramatic consequences for other branches of science and for technology. Physicist John Bardeen noted that "quantum theory opened up the possibility of understanding the properties of solids from their atomic and electronic structure," which led him and his colleagues at Bell Laboratories to the invention of the transistor and related devices. The revolution in electronics that followed brought new applications in computers, medical electronics, industrial controls, and communications that would have been impractical or impossible with the vacuum tubes that transistors replaced. Quantum mechanics also turned out to be essential for understanding basic chemistry, the properties of materials, molecular biology, and many other aspects of the physical world.

Today's experiments in elementary-particle physics can investigate phenomena 10^{12} times smaller than Thomson's. Yet, thanks to many ingenious advances in instrumentation, the actual scale of the largest modern experiments is only 10,000 times greater. Most of the experiments are conducted at a few large accelerator laboratories in the United States, Europe, and Asia, although some researchers obtain data from other sources, such as very high energy cosmic rays from outer space. Many different investigations can be conducted using a single large detector; although the detector project is frequently referred to as one "experiment," in essence a large detector is comparable to a whole laboratory in other fields of science. A decade or more can be spent on a detector's design, construction, use, and improvement. Almost all elementary-particle physics detectors are designed, built, and operated by groups that involve more than one institution; a typical group includes university faculty members and their students, accelerator laboratory staff members, postdoctoral researchers, engineers, and technicians. Collaborations range in size from 30 to more than 1,000 people; most are now international.

Creative problem solving is called for at almost every stage of both the experimental and the theoretical sides of elementary-particle physics. Elementary-particle physicists are intimately involved in the design and construction of their tools. Graduate students and postdoctoral researchers have the opportunity to master—and help develop—new approaches to integrated circuit design and fabrication, new algorithms and software, new techniques in precision engineering. Young researchers frequently devote most of their waking hours to their work; as in medical school, this total immersion is an important and valuable aspect of their training. Their experience in creatively solving novel problems, working with sophisticated technologies, discerning patterns hidden in massive data sets, and collaborating on large, complex projects is invaluable, whether they remain in elementary-particle physics or go into other fields, as more than half of them now do.

Modern elementary-particle physics experiments have enabled physicists to test theories that predict the behavior of elementary particles under extreme conditions, which sheds light on how the universe itself behaved in its earliest moments of existence. These cosmological insights, in turn, have brought physics to the point where the specific questions that are next on the elementary-particle physics agenda have the potential to illuminate some profound general questions—such as, What is matter? and What is force?—questions that only a few decades ago would have belonged to the realm of philosophy, rather than to experimental science.

Physicists expect that the next generation of experiments, which will be conducted with more powerful instruments than ever before, will reveal new phenomena crucial for understanding the origins of essential quantities, such as the mass of the fundamental particles, that at present can only be measured but not explained.

Although many of the specific questions that are ripe for investigation require fairly technical discussions to explain, the more general issues can be appreciated without specialized knowledge. These issues include the following:

• *Why are there three generations of elementary particles, and what accounts for their seemingly arbitrary progression of masses?* In addition to *u* and *d* quarks and the electron, one more elementary particle plays a role in our everyday lives: the electron neutrino. These four particles form a complete generation whose interactions can be described with precision by the theory that is now universally accepted by elementary-particle physicists. However, for reasons particle physicists do not yet fully understand, nature has been generous. There are two more complete generations of elementary particles, each analogous to the first one. They obey exactly the same principles as the particles in the first generation. They differ only in the masses of the analogues of the electron and the *u* and *d* quarks. For example, one of the electron's counterparts has 206 times the mass of the electron; the other has a mass 3,640 times the electron's. In a recent experimental triumph, the heaviest particle to be observed in the third generation, the *t* quark, weighed in with a mass about 60,000 times greater than its counterpart, the *u* quark. Theorists have proposed specific physical processes to explain the origin of these particles' masses; these ideas will be tested in the next generation of experiments.

• *Can Einstein's dream of unifying the known forces be realized?* Phenomena governed by the strong, weak, and electromagnetic forces can be described by a unified mathematical theory, but for many years no one could see how to include gravity in the description. Today, however, one of the most exciting areas of theoretical physics is an approach that would unify all four forces. It is called string theory, and some believe it represents a scientific revolution on the scale of quantum mechanics. Experiments at existing and planned accelerators will search for phenomena that are expected if string theory is correct, such as the existence of supersymmetric particles and the Higgs boson.

• *Why is there apparently more matter than antimatter?* Every known type of particle has an antiparticle counterpart, with the same mass and opposite electric charge. (Neutral particles either are their own antiparticles [e.g., the photon and the neutral pion], or have distinct antiparticles [e.g., the neutron and the antineutron].) When a particle and its antiparticle come close together, they are annihilated. Generally, whenever matter is created, an equal amount of antimatter is also created, so one would expect matter and antimatter to have been present in equal amounts in the early universe. If that were true, however, the universe should now be an excruciatingly dull place, since almost all pairs of matter and antimatter particles would have had more than enough time to encounter and annihilate each other. Why is there so much matter around—in the form of galaxies, solar systems, planets, and people?

Of course, as in any branch of science, serendipity and unforeseen developments are bound to play a key role in shaping the course of this work. Just as the Hubble Space Telescope is used to study many different phenomena, not all of which were even known when it was being built, particle accelerators and detectors are used to investigate issues that are recognized or become amenable to experiment only after the instruments are running. Elementary-particle theorist Steven Weinberg observed recently that physicists frequently "do not know in advance what are the right questions to ask, and we often do not find out until we are close to an answer."

Whatever future research in elementary particle physics reveals about the world around us, one thing is certain: It will inspire awe for the intrinsic beauty of the fundamental principles that shape our universe.

The following chapters report on the field of elementary-particle physics in a way that we think is accessible to readers without scientific backgrounds. Chapters 2, 3, and 4 present a comprehensive picture of the scientific status of the field today and how it reached this point. Chapters 5, 6, and 7 describe the research objectives and instruments for the next two decades. Chapters 8, 9, and 10 describe the structure of the field and how it relates to other branches of physics and technology and to society at large.

Finally, Chapter 11 presents the committee's conclusions concerning the health of elementary-particle physics and its recommendations for the future.

2

❖

What Is Elementary-Particle Physics?

INTRODUCTION

In the literal sense, nothing is simpler than an elementary particle: By definition, a particle is considered to be elementary only if there is no evidence that it is made up of smaller constituents. Yet, identifying the elementary particles, understanding their properties, and studying their interactions are turning out to be the key to illuminating why that most unelementary entity—the entire universe—is the way it is, how it came to be this way, and what its ultimate fate will be.

Philosophers through history have speculated on what matter is composed of. About a hundred years ago, atoms were considered elementary, until physicists learned that atoms consisted of electrons orbiting a nucleus. We now know that quarks make up the protons and neutrons inside the nucleus.

These particles are infinitesimal. Their scale is less than 10^{-18} cm—smaller relative to a grain of sand than a grain of sand is to the entire planet. Performing experiments to investigate the physics of elementary particles requires extremely sophisticated instruments and theoretical tools.

This chapter relives the ancient quest to find the fundamental constituents of matter, shows how this quest is shaped by relativity and quantum mechanics, introduces the forces (interactions) among particles, shows what we know about how orderly the universe is, and describes the technique of particle collisions that has revealed so much about its inner structure and beauty.

FUNDAMENTAL CONSTITUENTS OF MATTER

For centuries, philosophers have asked, "Are there a small set of fundamental constituents out of which everything is made? Or might it be that we will always find structure within structure, layers within layers like an onion?" Thales of Miletus, who suggested that water was the single fundamental entity from which matter is built, may be the first person recorded to suggest the appealing notion of an ultimate form of matter. The following material briefly sketches how this thinking has progressed to the present time.

Earth, Air, Fire, and Water

Thales' student Anaximander added earth, fire, and air to water to the list of fundamental building blocks. The important notion that *rational* principles could explain what was observed was contained in this philosophy.

Chemical Elements, the Periodic Table, and Atoms

What are earth, air, fire, and water made of? Addressing this question led, by the nineteenth century, to recognition of the familiar chemical elements. The definition adopted then for an "element" was a substance that cannot be decomposed further into simpler substances by ordinary chemical means. Thus, the world consisted of a number of distinct substances (at the time, only about 30 elements were identified; today, there are more than 100). It was discovered that elements combine with other elements according to very simple rules: Two hydrogens plus one oxygen make one water "molecule." The essentially limitless number of chemical "compounds" that are found in nature are then the result of combinations of elements. At the beginning of the nineteenth century, John Dalton proposed an atomic theory of matter: The elements themselves are collections of tiny indestructible atoms characterized by their atomic weights (oxygen atoms weigh 16 times as much as hydrogen atoms). This can be considered the first theory of "elementary particles" having a sound scientific basis.

Dmitri Mendeleev arranged a table of the elements in order of increasing atomic weight and thereby made one of the most important discoveries in the history of science: The properties of the elements are "periodic" functions of their atomic weights. For example, in Mendeleev's "periodic table," the metals copper, silver, and gold, which have vastly different atomic weights, all appear in the same column. There were also gaps in this table: Elements not yet discovered that should exist if this atomic theory were correct. Confirmation of Mendeleev's "predictions" then made this scheme the basis of chemical thinking in the late nineteenth century; by the early twentieth century, further experiments established atoms as real physical entities.

Protons, Neutrons, and the Electron

Although atoms were thought to be elementary, they too are composite objects. The atom, as first revealed in experiments by Rutherford, is an electrically neutral object, approximately 10 billionths of a centimeter in diameter. This means that there are about 2 million atoms stretching across the diameter of the period at the end of this sentence. Each atom has a tiny positively charged nucleus, about 10,000 times smaller yet. Negatively charged electrons occupy the space surrounding the nucleus. The mass of the electron is about 2,000 times lighter than the mass of the hydrogen nucleus: the proton. Electrons were the first of the modern elementary particles to be discovered.

The mass of most nuclei is about twice the mass of the protons they contain. The additional mass is provided by another particle, the neutron, which has a mass very close to that of the proton but is electrically neutral.

Today's Fundamental Constituents

It is natural to ask what protons, neutrons, and electrons are made of. With today's particle accelerators, one can "look inside" these objects for an internal structure. The proton and neutron are found to be made up of "quarks."

Two quarks with a positive electric charge of 2/3 (of the electron charge), called u quarks, and one quark with a negative electric charge of $-1/3$, called the d quark, make up the proton. Similarly, the neutron is made up of one u quark and two d quarks. There are many other particles that can be built out of the quarks combined in particular ways; these are called hadrons.

Physicists have also tried to see if there is anything smaller inside the electron. Experiments have the sensitivity to detect objects even 10,000 times smaller than the proton itself, but nothing has been found.

As far as physicists today know, quarks are also fundamental and are not made of yet smaller constituents. The question is still open experimentally, but theory and experiment are pointing more than ever before to the possibility that we have discovered the "ultimate constituents."

The Neutrino and Leptons

There is a fundamental particle, called an electron neutrino, that does not combine with other particles in the way that quarks combine to make hadrons, hadrons combine to make nuclei, and electrons and nuclei combine to make atoms. This is a massless or almost massless particle that carries no charge and is, as shown later, a "partner" to the electron, as its name implies. These are the first members of a class of particles different from quarks, which are called leptons: the electron and its associated neutrino.

Particle "Generations"

So far then, a charge 2/3 quark, a charge −1/3 quark, a neutral lepton, and a charge −1 lepton have been discussed. The masses of these particles in units of 10^9 electron volts (GeV) are shown in the first part of Table 2.1; and they comprise what is called a particle generation or family. A major surprise has been production in the laboratory of extra particle generations. A remarkable feature of nature that has been discovered is that this pattern of particles—two quarks and two leptons of the indicated charges—is repeated and then repeated again. Except for the neutrinos, which perhaps remain massless, the particles of each subsequent generation become heavier, as Table 2.1 shows. These additional generations appear to have nothing to do with "ordinary tangible matter." Yet they were important in the first moments of the universe and have a profound role in our understanding of nature.

The masses of the quarks and leptons range from zero, or near zero, for neutrinos to almost 200 times the proton mass for a *t* quark. Understanding why quarks and leptons exhibit this not quite random progression of masses is an important topic of research in elementary-particle physics.

Good experimental evidence exists that there are only three generations. Why this should be so constitutes a major mystery in the field.

TABLE 2.1   Particles of All the Generations

Particle	Mass (GeV)	Electric Charge
First generation		
u quark	0.005	+2/3
d quark	0.008	−1/3
electron neutrino	0	0
electron	0.0005	−1
Second generation		
c quark	1.2	+2/3
s quark	0.175	−1/3
muon neutrino	0	0
muon	0.106	−1
Third generation		
t quark	175	+2/3
b quark	4.25	−1/3
tau neutrino	0	0
tau	1.781	−1

TABLE 2.2   Major Operating U.S. Accelerator Laboratories

Laboratory	Primary Particle	Energy (GeV)	Location
Fermilab	Proton	1,000	Batavia, Illinois
Stanford Linear Accelerator	Electron	50	Palo Alto, California
Brookhaven	Proton	30	Upton, New York
Cornell Electron Accelerator	Electron	10	Ithaca, New York

RELATIVITY, QUANTUM MECHANICS, AND PARTICLE ACCELERATORS

The two major scientific revolutions of the twentieth century—relativity and quantum mechanics—still provide the basic framework for describing elementary-particle physics.

Quantum mechanics tells us that particles have wave-like properties. These are not observed for large objects such as billiard or baseballs, but for particles with small masses the wave nature becomes evident. In quantum mechanics, the wavelength of a particle is inversely proportional to its momentum. This means that as the momentum (and therefore the kinetic energy) of a particle increases, its wavelength decreases. This is the reason high particle energies are required to probe small distances and is the prime motivation for use of the high-energy particle beams produced at particle accelerators. Chapter 7 discusses just how these technological marvels work; for now, Table 2.2 lists the major accelerators presently used in particle physics in the United States.

With high-energy accelerators, particle physicists can effectively "trade" energy for mass, allowing them to directly produce particles that weigh many times more than the particles being accelerated. This follows from relativity, which says that a particle with mass m that is at rest has an energy E given by the famous equation $E = mc^2$. Thus, if two protons each having an energy of 1,000 GeV can be brought together, it would in principle be possible to produce in such collisions two new particles (at rest) each weighing 1,000 GeV, or about 1,000 times as much as the initial protons. This is the means by which very heavy members of subsequent particle generations were discovered. It is analogous to the collision of two tennis balls to produce a bowling ball. The analogy would be even more accurate if this were the *only* way to produce and observe a bowling ball!

More implications of relativity and quantum mechanics, important for particle physics, can be indicated with a discussion of one of the particles under intense study: the *B* meson.

First, the *B* meson consists of a *b* quark and an anti-*d* quark. Antimatter was deduced in the 1930s by attempts to understand how the motion of electrons was defined by quantum mechanics and relativity together. In the equation for the electron developed by P.A.M. Dirac, there appeared an extra solution having opposite charge to the electron; this turned out to correspond to the *positron*, the antiparticle of the electron. This prediction was a brilliant deduction whereby an entire and formerly hidden sector of the world was uncovered using only mathematics and reason! However, mathematics includes many possibilities that are never realized in nature, and this "antiworld" had to be verified by experiment. Physicists now know that every type of particle has a corresponding antiparticle, a symmetry that effectively doubles the number of types of particles in nature (except for the kinds of particle that are their own antiparticle). Antimatter played a major role in the evolution of the early universe, but as shown in the previous chapter, a key question in particle physics is why the universe today appears to be made of matter only.

Second, the *B* meson is unstable, having a mean life of approximately a trillionth of a second. In fact, most types of particles are unstable: Even in a total vacuum, they spontaneously disintegrate or decay into lighter particles. How does the *B* meson decay? The mechanism involves the possibility of a *B* meson directly turning into an important particle that has almost 20 times its energy—the *W* particle—and the *W* then rapidly decaying into lighter particles. (The important role of the *W* is discussed below.) It is the Heisenberg uncertainty principle of quantum mechanics that permits these momentary extreme violations of the conservation of energy. Such processes are called "virtual," since they cannot be directly detected. However, through virtual processes, effects that would otherwise be expected to be seen only at very high energies can be detected (albeit infrequently) at much lower energies.

Third, we cannot know exactly when a *B* particle will decay. Quantum mechanics provides a precise expression for the *probability* of a decay at a particular time; this probability is all we can know. It is still not known why the world obeys quantum mechanics, but that it does is both beautiful and incontrovertible.

Finally, a particle's decay time will depend on its speed. The inner workings of the *B* particle, as Einstein taught, slow down significantly the faster it travels. This effect ("time dilation") makes it possible for particle physicists to directly study short-lived new particles by extending their lifetimes in the laboratory frame so that they travel further in a detector. An example of this is shown in the cover illustration to this volume.

FORCES

In addition to identifying elementary particles, physicists try to understand the means by which they interact. At the small scales encountered in high-energy physics, these forces, or interactions, can also create and destroy particles. It appears that only four forces are needed to describe the behavior of all matter in the universe. Their characteristics are described below.

Gravity

The force of gravity is the most familiar one in our everyday experience. A chain of observations and reasoning by Galileo, Brahe, Kepler, and Newton led to the universal law of gravitation. This law describes a force between two masses that is always attractive, proportional to the product of the masses of the two objects, and inversely proportional to the square of the distance between the objects. It accurately describes the motion of falling objects near Earth's surface and, amazingly enough, accurately applies to the motions of celestial bodies. As such, Newton's law of gravitation is a "universal" law. More than 200 years later, Einstein's theory of general relativity subsumed Newton's law and predicted small deviations, which were confirmed.

Acting for billions of years on the galaxies, gravity is the force that has had a profound effect on the structure of the universe, second only to whatever interaction initially produced the matter out of which the universe is made.

Electric and Magnetic Forces; Electromagnetism

Electricity

The electric force accounts for all everyday phenomena that are not gravitational. The muscular processes in our bodies, tension and compression in the structural members of tall skyscrapers, the combustion of chemical fuels to power our society: All are examples of the electric force. Much of modern technology utilizes the electric force: The motion of charges in electronic circuits, television screens, and computer monitors—all rely on the electric force.

Unlike gravity, the electric force is both attractive and repulsive. The atom is electrically neutral (with the positive charge of the nucleus balanced by the electrons surrounding it), but the arrangement of its electrons determines its chemical activity. It may be an inert noble gas, it may combine in a regular way to form a crystalline metal or a complex molecule, and complex molecules may combine to form living cells. In all of these cases the electric force is at work.

Magnetism

Magnetism used to be considered the third macroscopic force. Two bar magnets exert a force on each other that is attractive if the bars are oriented one way and repulsive if one of the bars is reversed.

One of the most far-reaching phenomena in physics was discovered by Hans Christian Oersted in 1820 when he observed that a magnetic compass needle was deflected when brought near an electrical current. Eleven years later, Michael Faraday and Joseph Henry independently discovered that a moving bar magnet produces an electrical current in a nearby conductor. These experiments strongly suggested that electricity and magnetism were deeply interrelated.

A consistent description of both phenomena was first achieved by James Clerk Maxwell. Interestingly, his equations also implied that electromagnetic "radiation" could travel in vacuum. His equations predicted the speed of this radiation, which turned out to be the speed of light. Heinrich Hertz in a classic experiment was able to create this electromagnetic radiation and detect it a distance away. Thus, light itself was understood for the first time to be propagating electromagnetic energy, and the first step was taken in the development of our modern telecommunications industry. This huge industry is just one of many made possible by the application of the theory of electromagnetism as described by Maxwell's equations.

Weak and Strong Forces

There are two forces that apparently have little significance in everyday life because they operate only at subatomic distances. Nevertheless, they play crucial roles in physical processes important to us. These are the weak force and the strong force.

The Weak Force

The "weak" force between elementary particles is much weaker than electromagnetic forces. It is a very short-range force, acting only over microscopic distances (10^{-15} cm).

The weak force controls the nuclear fusion reactions by which the Sun and stars shine. Deep within our Sun, the density is so great and the temperature so high that nuclei can overcome the repulsion from the electrical force and release energy by fusing together. Neutrinos created by this weak interaction carry energy out of the star to cool it, controlling its temperature (and consequently that of its surrounding planets). In other reactions, photons carrying electromagnetic energy are emitted. It is of course the photons emitted by the Sun that warm Earth's surface and help to sustain life.

Earth also has an internal source of energy, which offsets the heat being radiated back into space from its surface. This is supplied by radioactive decay of heavy nuclei in Earth's interior. One of the key processes in radioactive decay of heavy nuclei is called beta decay—another manifestation of the weak force—whereby a neutron in a nucleus transforms into a proton and the nucleus emits an electron and a neutrino.

The Strong Force

The protons within the atomic nucleus repel each other through the electric force, but the nucleus does not fly apart. It is bound together by another very short-range force, the strong force. The stability of all the matter in our everyday experience comes about through the action of the strong force. It is also the strong force that binds together quarks inside the proton and the neutron. An interesting and important aspect of the strong force between quarks is that the strength of the force increases as the quarks get further apart, somewhat like stretching a rubber band tightly. This explains why quarks are never seen in isolation; they are said to be "confined."

A relatively large amount of energy is stored in heavy atomic nuclei, in the form of the energy it takes to hold protons and neutrons together. Heavy nuclei, such as uranium-235, when bombarded by neutrons become unstable and split into lighter products (nuclear fission), releasing a great deal of this energy. In a nuclear reaction, about a million times more energy is released than in a typical chemical reaction such as the burning of carbon. This energy, resulting from the strong force, is the source of nuclear power.

What "Transmits" Forces?

The four fundamental forces encountered—gravity, electromagnetism, the weak force, and the strong force—underlie all observed phenomena. Over the years, other forces have been hypothesized, but experiments searching for them have so far produced null results. More sensitive experiments in the future might discover still other forces.

Physicists understand today that when quantum mechanics and relativity are taken into account, forces actually arise from the exchange of other particles. Physicists then speak of two distinct types of particles: matter particles and particles that carry forces.

The electromagnetic force is communicated between particles via exchanges of photons. These photons are the same quanta of energy that are familiar to us as radio waves, light, and x rays. The carriers of the weak force are W and Z bosons, first detected directly in 1983. (The interesting effects of the W boson were first seen in the observation of radioactive decay at the end of the last century, and it took about 85 years for a W boson to be directly produced in the laboratory.) The carriers of the strong force are called gluons. Gluons have not been observed in isolation—they, like quarks, are confined—but direct evidence for their existence is seen routinely in experiments. Also, gluons are present in hadrons and can be considered a constituent of, for example, the proton.

The carrier of the gravitational force has been named the graviton, but it has yet to be directly observed, because the gravitational force is exceedingly weak.

TABLE 2.3   The Four Fundamental Forces

Force	Force Carrier	Approximate Mass of Carrier
Gravity	Graviton	0
Electromagnetism	Photon	0
Weak force	W and Z bosons	About 90 times the proton mass
Strong force	Gluon	0

Table 2.3 lists the force carriers. Note that although three of them are massless, the fourth is extremely heavy. Given this disparity, it might seem impossible that they could arise from a single mechanism. Yet, as the next section indicates, physicists are optimistic that all the forces can be described in a "unified" framework.

Unification of Forces

The question as to whether some or all of the observed forces have a common origin ("unification") is a major theme today in particle physics. The first great example of unification goes back to Newton, who realized that the motion of rocks (or apples) falling toward Earth was governed by exactly the same law that governs the motion of the Moon around Earth or the attraction of one star by another. As shown earlier, electricity and magnetism, two forces responsible for dissimilar physical phenomena, are actually different manifestations of a single force—electromagnetism. Part of this unification was the profound realization that light was nothing more than a propagating electromagnetic wave.

Further unification is treated in the following chapter. It has been established that two of the forces listed in Table 2.3 (electromagnetic and weak) are indeed unified, and there is compelling evidence that a third (strong) is as well. It may even be that gravity is also subject to unification so there would be in a real sense just one force.

LAWS OF NATURE

Underlying the electrical, magnetic, gravitational, and other phenomena of particle physics are a number of general principles. Examples of the most fundamental of these principles (often referred to as laws) are the conservation of energy, the conservation of charge, and the law of cause and effect (causality). Physicists believe these laws to be universal and absolute, applying to interactions between the smallest components of matter that have thus far been observed, to interactions between galaxies of stars, and even to the development of the universe itself.

Our universe is not "random": The behavior of atoms and the laws of quantum mechanics that physicists study here on Earth are found to hold everywhere else. By studying light, radio waves, and other electromagnetic energy from the most distant stars, physicists can determine that the forces and rules discovered on Earth apply equally well there, far away and at much earlier times in the evolution of the universe. As discussed in the next chapter, physicists have realized that the conservation of energy is a direct consequence of the fact that to a good approximation the laws of physics do not depend on time. It did not have to be so, and this observation is of profound significance.

PARTICLE COLLISIONS

How do particle physicists uncover the phenomena that are described? The large accelerator laboratories have been mentioned already. This section briefly characterizes what goes on at such laboratories where the key study is of collisions of particles; Chapter 6 elaborates more fully.

Scattering Experiments

Perhaps the archetypal example of probing the subatomic world involved experiments (alluded to earlier) performed by Rutherford from 1909 to 1911. He and his colleagues directed a beam of alpha particles (nuclei of helium), which originated from radioactive decay, at a thin gold foil. The atoms of gold were transparent to the bombarding particles. Occasionally, however, some of the alpha particles were scattered backward from whence they came, as if they had encountered an object with much greater mass (the atomic nucleus) than their own. In a memorable statement after making these observations, Rutherford said, "It was almost as incredible as if you fired a 15-inch shell at a piece of tissue paper and it came back and hit you!"

The principle of directing beams of subatomic particles to collide with other particles, observing what emerges from these encounters, and interpreting the results via models of their interaction remains the major technique for exploring the physics of elementary particles. Accelerators provided more energetic beams that were then used to study phenomena at much smaller distances than could be done simply by using the particles from natural radioactivity. By this technique, new heavier particles are produced. Also, intense beams of new particles can be made and the ways in which they decay or scatter on other targets can be studied.

Colliders

For the study of phenomena revealed only at the very highest energies (e.g., production of the very heavy t quark or of W and Z bosons), the technique of "colliding beams" is employed. Here, one beam is directed at another rather than at a fixed target. In collisions of billiard balls, for example, the effective collision-energy increase is a factor of four if two balls of the same speed collide head-on as opposed to one of them being at rest. However, because of relativity and the conservation of momentum, the effective energy of the collision of particles aimed at each other is far greater. At the Fermilab Tevatron, collisions have been made between 900-GeV protons and 900-GeV antiprotons. Achieving the same collision energy with a stationary target would require an accelerator with a circumference about 2,000 times as large as Fermilab's, or about 8,000 miles!

There are different types of colliders operating today that are important for particle physics. The relatively large mass of protons and antiprotons makes it more efficient to accelerate them to high energies than to accelerate electrons or positrons. However, in the collisions, it is their constituents—quarks and gluons—that interact. Since many subatomic constituents make up the proton and antiproton, no single one carries the full energy of the accelerated particle. For example, in proton-proton collisions, the *effective* collision energy is about a factor of 10 lower than the full energy of the beam. By increasing the intensity of the beams, it is sometimes possible to study higher-energy processes.

In collisions between electrons and positrons, the energy of collision is the full energy to which the particles are raised (since electrons and positrons appear to have no substructure). Such collisions cleanly probe the electromagnetic and weak interactions: They do not create the extraneous debris characteristic of proton collisions and are easier to interpret.

Finally, because an electron lacks substructure and behaves in a point-like way, it is a useful probe for exploring the structure of the proton. An electron-proton collider provides information about the structure of a proton that is not available from a proton collider and provides an opportunity to search for hypothesized objects that combine both quark and lepton characteristics.

SUMMARY

The variety of phenomena that particle physicists have uncovered and are studying has been surveyed in this chapter. Although these phenomena occur at the smallest distance scales, what is observed has relevance in understanding the physics of forces that govern the atom, the energetic processes in cores of stars, and even the structure of the universe.

The collisions of high-energy particles have been shown to reveal new and important structure. These collisions re-create the conditions of the universe just after its birth. The laws that are discovered have existed for all time and everywhere in the universe. A small number of forces have been discovered, all of which could arise from a single one. The particles on which these forces act have a mysterious structure; the lighter ones make up our everyday world, whereas the role of the next two generations still represents a major puzzle.

The following chapter presents the theoretical framework in which these phenomena are currently understood.

3

❖

Symmetries, Forces, and Particles

INTRODUCTION

Several decades of experiments in high-energy physics have produced an extremely rich and detailed set of data on elementary particles, the most important of which are summarized in the 700-page *Particle Data Book* (*Physical Review D*, vol. 50, no. 3, part 1, August 1994: Web site at http://pdg.lbl.gov/pdg.html). For example, the particle known as the *K* meson is unstable and has about 70 ways in which it can disintegrate into a set of lighter particles. For each decay mode, many measurements have been made—for example, the nature, speeds, and directions of the particles produced. How can one make sense out of this wealth of data? Theoretical particle physics aims to uncover the regularities hidden in the data and to formulate rules and laws that provide an understanding of the data in as simple and precise a way as possible.

The most useful tool in this enterprise is *symmetry*, which is described below. Symmetries have allowed an extremely compact synthesis to be made of all of the knowledge of particle physics: the Standard Model. There are now many very precise tests of this synthesis. A description of how the strong, weak, and electromagnetic forces arise as a consequence of three symmetries can be found below; indeed, the symmetries dictate precisely the form of these three interactions.

Note: This chapter presents a picture of the rich theoretical underpinnings to the field of elementary-particle physics. Although it attempts to do so without a great deal of technical language, some readers may find parts of this chapter more challenging than the rest of the report.

However, these symmetries lead to an astonishing—and obviously wrong—prediction: All of the elementary particles should be massless. To reconcile this puzzle, the symmetries must be "broken," which implies that there must be new forces that have not yet been discovered. In the Standard Model, these new forces are related to a hypothetical particle called the Higgs boson. Interaction of the Higgs boson with other particles generates particle masses but does not provide real understanding of the observed pattern of masses. Also, one might expect such a theory to generate particle masses that are much heavier than observed. The masses of elementary particles are a crucial clue in deciphering the ultimate theory of nature, and much detective work lies ahead.

With all of its successes in giving order to the wealth of data from high-energy particle collisions, the Standard Model brings a whole new set of questions into sharp focus. What determines the particles, symmetries, and mass scales of this theory? Could they have been very different, completely changing the nature of the world in which we live? Physicists are able to describe the physical universe with astonishing simplicity and precision but have very little understanding as to why it is this way. Several theoretical proposals are discussed later that extend the Standard Model to address, particularly, the question of particle masses. There are other issues that even these theories do not begin to address, such as the role of gravity. Superstring theory, which offers the hope of a complete, all-encompassing theory of the origin of particles—together with all of their symmetries and interactions—is discussed at the end of this chapter.

SYMMETRIES

Symmetry arguments have a long and honorable history in physics, but only in recent times have they come to dominate our understanding of fundamental physics. The power and beauty of such arguments became fully apparent only when expressed in mathematical form. The committee hopes, however, to convey the spirit of this subject by discussing some of its central physical concepts.

What precisely do we mean when we say that physical laws display symmetry? A square is symmetric when rotated through 90 degrees around its center. This operation produces an orientation identical to the initial one—we say that a symmetry operation leaves an object invariant. A circle is left unchanged (invariant) by a rotation through any angle. Since it allows more symmetry operations, it possesses a larger symmetry. More generally, an object is said to be symmetrical when there are operations on the object that could have changed its appearance but in fact do not.

Similarly, physical laws also have symmetry. One of the basic principles of physics is that the laws of physics at one location are the same as at another, and at one time, the same as at another. This principle is equivalent to a symmetry: The laws of physics are invariant when we change our viewpoint—either from one location to another or from one time to another.

Rotations provide another familiar example of a symmetry. Imagine a laboratory as a windowless spaceship in free-fall, isolated from electric and magnetic forces, in which an experimentalist has some apparatus to determine a certain law of physics. Suppose that the experimentalist makes a measurement, the spaceship is rotated, and the same measurement is made again. The results of the two measurements are always found to be the same: When writing the laws of physics, there is no need to specify the orientation of the laboratory. Much less familiar, and more far reaching, it is not necessary to specify the speed of the laboratory—the laws of physics do not change if the experiment is done, for example, on an airplane. This last symmetry principle was the crucial one that led Einstein to develop special relativity. All of the astonishing results of special relativity—such as the equivalence of mass and energy and the inability to travel faster than light—follow from the requirement that physics be symmetrical under the operations discussed.

Symmetries and Particle Physics

How are these symmetries of space and time—which are called space-time symmetries—relevant to particles and their interactions? First, there is a direct consequence for the very nature of elementary particles themselves. Rotational symmetry leads to elementary particles' possessing a new attribute, called spin. For example, electrons come in two varieties: left-handed and right-handed. The difference can be pictured in terms of the view of a football's spin as seen by the quarterback who threw the pass. The football spins clockwise if thrown by a right-hander and counterclockwise if thrown by a left-hander. Photons also come in the same two varieties of spin, but other particles have three spin orientations, and still others are spinless.

The symmetries of space and time also constrain the rules by which all particles interact. One should not forget why it is so important to understand these interactions. The explanation of every physical process, from the growth of plants to space shuttle lift-off, has its fundamental origin in these interactions. The properties of materials, from concrete to quicksand, ultimately depend on the properties of elementary particles.

Consider a collision between two electrons, shown in Figure 3.1. The two electrons labeled 1 and 2 approach each other, collide, and then leave as the two electrons labeled 3 and 4. Quantum mechanics says that there is no unique outcome: Sometimes the electrons are deflected by large angles, sometimes by small angles. In this quantum world the laws of physics can be phrased in terms of probabilities. If the laws determine the probabilities for all possible outcomes, then the description is complete. The problem is that the probabilities depend on the speeds, directions, and spins of each of the four electrons for which there are an infinite number of possible values. The importance of symmetry can now be appreciated: Probabilities depend on the speeds, directions,

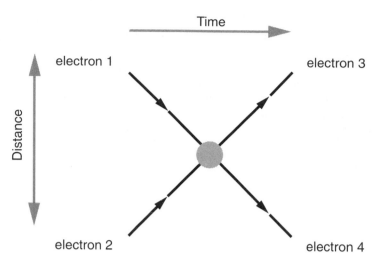

FIGURE 3.1 Collision of two incoming electrons, 1 and 2, into two outgoing electrons, 3 and 4.

and spins of particles in a way that is determined by symmetries. In general, *symmetries place powerful restrictions on the nature of interactions between elementary particles.* Probabilities can be expressed in terms of just a few numbers (called parameters), even though there are many possible initial and final configurations.

As physicists discovered more elementary particles, they found that patterns in their properties could be understood in terms of mathematical symmetries. These newer symmetries often act in more abstract, so-called internal spaces. For example, early experiments on protons and neutrons revealed that although their electric charges differ, these particles are fundamentally similar and their strongest interactions are identical. This close similarity inspired the concept of a two-dimensional "internal" space, in which protons and neutrons correspond to different directions. The similarity of their behavior becomes the statement that physics is unchanged as one rotates in this imaginary internal space.

If the laws of physics are not changed by such a rotation, this operation is said to be an internal symmetry. The fruitfulness of this way of thinking was revealed as additional particles were discovered. Each had to be located in internal space. Apparent asymmetries in the distribution of known particles within internal space would appear if one had only part of the picture. In this way the existence of particles necessary to complete the symmetric pattern, and some of their properties, could be predicted. An early example of this was provided by discovery of the *omega*-minus (Ω^-) particle, which was predicted ahead of time to fill a gap in the symmetrical pattern shown in Figure 3.2. The

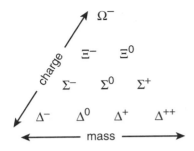

FIGURE 3.2 Pattern of particles that allowed prediction of the Ω^- particle. Particles in the same row have similar masses; particles with the same electric charge (shown by superscripts) also lie on straight lines.

most recent example is the discovery of the *t* quark, which completed the particles listed in Table 2.1.

Local or Gauge Symmetries

The most profound and powerful symmetries of physical law established at present are so-called local or gauge symmetries. Such symmetries underlie both Einstein's theory of gravity (also called general relativity) and the Standard Model of strong, weak, and electromagnetic interactions. Most of the matter on Earth is made up of just two quarks, the so-called up and down quarks, denoted by *u* and *d*. (A glossary to describe many of these terms is included in the Appendix at the end of this report.) Because of internal symmetries, each of these quarks comes in three varieties, which are labeled by colors: u_r, u_g, and u_b represent the red, green, and blue up quarks. Physical laws are invariant if quark colors are interchanged, for example, if u_r and u_g are switched. In fact, a local symmetry means that physical laws are unchanged even when different interchanges are made at different locations in space. For example, one might switch u_r with u_g in one part of the laboratory and u_r with u_b in another part. There are an infinite number of such local operations, and requiring the laws to be unchanged under any of them is extremely constraining.

Local internal symmetries actually *require* the existence of particles (called force carriers) whose interactions are the origin of the forces. The local symmetry that acts on the three colors leads to the strong force that binds quarks into nuclei. Insight into this basis for the understanding of forces can be gained by returning to the example of the collision of two electrons in Figure 3.1. The circular blob represents the actual interaction between particles and is highly constrained if it arises in a theory with a local internal symmetry. If one could look inside this blob at high magnification, such a theory would dictate that the interaction results from exchange of a force particle, which for an electromagnetic interaction is called the photon, as illustrated in Figure 3.3. Furthermore, the interaction of a photon with two electrons is itself greatly constrained by local symmetry. No matter what the speeds, directions, and spins of the par-

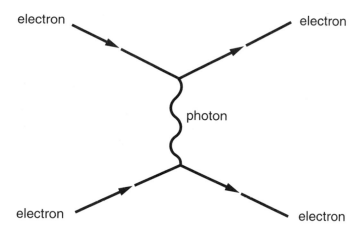

FIGURE 3.3 Collision of two electrons resulting from exchange of a photon.

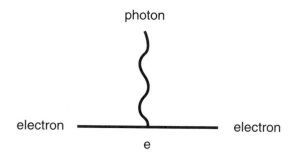

FIGURE 3.4 Electromagnetic vertex of an electron: The electron emits a photon with a probability, in emissions per second, proportional to the electron's charge.

ticles, only a single parameter—the electric charge of the electron—is needed to describe this interaction. This is the same single parameter that enters all electromagnetic interactions of the electron, for example, the bending of the path of an electron in a magnetic field and the electron's binding to an atomic nucleus. The electromagnetic interaction is represented diagramatically in Figure 3.4 as a vertex at which the three particles come together. (In these figures, straight lines represent matter particles and wavy lines represent force particles.) Local symmetries are also called gauge symmetries, and the resulting force particles, such as the photon, are known as gauge bosons.

It is startling to realize that the apparent infinite variety of chemical properties and reactions of atoms and molecules all result fundamentally from this single electromagnetic vertex. The very existence of the photon, as well as the form of the electric and magnetic interactions, is a consequence of the local electromagnetic symmetry.

It is clear that symmetries are the most powerful tool physicists have for understanding the properties and interactions of particles, yet only by careful experimentation can we learn which symmetries nature possesses. Many symmetries have been proposed, but measurements provide the only sure guide. Future experiments will continue the quest to uncover more of nature's symmetries, and theoretical physics will struggle further to understand why nature has chosen these symmetries.

THE STANDARD MODEL

A major development in theoretical physics this century was the construction of what are called quantum field theories—theories of particles and their interactions that incorporate the probabilistic laws of quantum mechanics, special relativity, and the symmetries discussed above. This enterprise began soon after the discovery of quantum mechanics in the late 1920s. The quantum field theory of electromagnetism, describing the electron and the photon, reached its final form in the late 1940s, but theories involving larger local internal symmetries were not fully understood until the early 1970s. Quantum field theories are the basic tool for theoretical particle physicists. There are many such theories, and the great variety of phenomena they can describe is the subject of continuing research.

The Standard Model is a quantum field theory that provides a concise and accurate description of all known particle phenomena. This discussion relies on the ideas of symmetry and interaction vertices introduced in the previous section.

Three local internal symmetries have been discovered in nature: They are called strong, weak, and electromagnetic, after the three forces to which they give rise.

Strong symmetry leads to force particles of the strong interactions—the gluons, g. The matter particles that feel this force are called up and down quarks, u and d, and come in red, green and blue varieties. The gluon vertex for the up quark is illustrated in Figure 3.5. A quark of one color goes into the interaction and comes out as a quark of a different color, but its other properties are not changed. The mathematical theory of quarks and gluons that underlies this vertex is called "quantum chromodynamics," or QCD for short. The strength of the gluon interaction is called g_3. It is large, making this QCD interaction strong. The matter particles that do not feel this strong force are leptons: the electron, e, and its neutrino, v_e, as well as their second- and third-generation counterparts, the muon and tau, and their respective neutrinos.

The force particles of the weak interaction, the W and Z bosons, are very massive, which results in a force with a very short range—much less than the

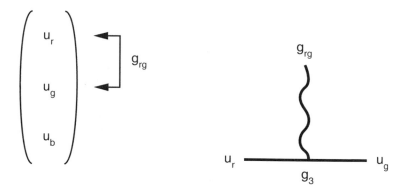

FIGURE 3.5 Quark triplets and the gluon vertex.

diameter of a proton. The electromagnetic interaction, on the other hand, has a massless force particle, the photon, with a corresponding range of interaction that is infinite, allowing us to see to the edge of the universe. A single parameter g_2 describes the strength of the weak interactions (see Figure 3.6), whereas g_1 gives the strength of the electromagnetic interaction.

The weak symmetry has a very peculiar property. Only counterclockwise-spinning (left-handed) quarks and leptons feel the weak force. The reason nature treats left-handed and right-handed objects differently is one of the many questions about the nature of forces for which we have as yet no adequate answers.

As indicated earlier, symmetries dictate both the forces and the so-called multiplet structure of particles that feel these forces. Table 3.1 summarizes this information about how the three forces that arise from local internal symmetries relate to the four basic types of matter particle (u, d, e, v_e). For strong and weak forces, the entries represent the size of the multiplet of particles that interacts with the corresponding force particles. The strong force acts among triplets of quarks (three colors), changing one into the other; the weak force acts between quark and lepton doublets, again changing one into the other. An entry "1" implies that there is no interaction, since there is nothing to change into. Electromagnetic force acts on all particles except neutrinos (not changing their nature), and the entry in Table 3.1 gives the electric charges of particles.

Table 3.1 represents the limit reached at present in the quest for a simple understanding of particle interactions. There are several questions that this knowledge raises, questions that are not answered by the Standard Model: Why is it that some particles feel the strong force and some do not? Why are the weak interactions left-handed? Why are there not multiplets having more than three components? In short, why are the matter particles what they are, and why do they interact with force particles in the way shown in Table 3.1? This table offers a mystifying array of numbers—how can it be understood?

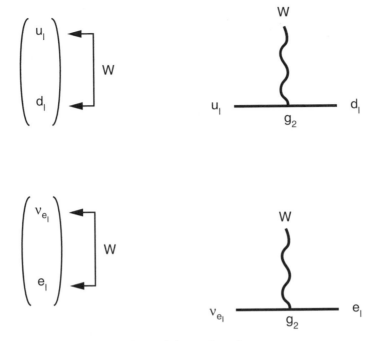

FIGURE 3.6 Left-handed doublets and the weak vertices.

TABLE 3.1 The Aperiodic Table

Force/Particle	(u_1, d_1)	u_r	d_r	(v_{e_1}, e_1)	e_r
Strong	3	3	3	1	1
Weak	2	1	1	2	1
Electromagnetic	2/3, −1/3	2/3	−1/3	0, −1	−1

The four matter particles discussed so far (u, d, e, v_e) are the members of the first family, or generation, of particles. Three such generations of particles have been found, as shown in Table 2.1. The only known difference between the three generations is their mass—in particular, the force particle vertices of the heavier generations are identical to those of Figures 3.4, 3.5, and 3.6 for the lightest generation. This replication of particles suggests to some that there is a new internal symmetry to be discovered that is responsible for the different generations. Physicists believe that some deeper understanding of the three rows, or periods, of Table 2.1 will eventually be found, in the same way that the revolution of quantum mechanics led to an understanding of the periodic table of the elements. In contrast, the particles of a single generation cannot be grouped into subgroups or periods of particles with similar properties—Table 3.1 does not have a periodic structure.

Spontaneous Symmetry Breaking

Whereas interactions of the force particles are restricted by the three local symmetries, the observed masses of the quarks are restricted by the strong symmetry. For example, although u_r, u_g, and u_b have the same mass, members of weak doublets, such as v_e and e, do not. The nonzero masses of elementary particles are said to break electroweak symmetries (i.e., if the electroweak symmetry was unbroken, the masses would all be zero). This seems unsatisfactory—surely all aspects of a theory should have the same symmetry. In fact, physicists believe that the equations of the theory do initially possess electroweak symmetries but that something within the theory causes the solutions to the equations to break the symmetry. This important phenomenon of spontaneous symmetry breaking can be illustrated by the examples of the square and circle mentioned earlier. Recall that these shapes are symmetrical when rotated about their centers, by 90 degrees for the square and any angle for the circle. If a square and a circle are drawn on an elastic sheet and the sheet is stretched in one direction so that these shapes are elongated into a rectangle and an oval, we have broken the symmetry. Now, if the rectangle is rotated by 90 degrees, it does not match the shape corresponding to its original position. This stretching is a simple analogy for the spontaneous symmetry breaking that occurs in theories of particle physics. The first step is to infer from data what stretching is occurring—which is understood quite well—and the next step is to understand what is causing the stretching. Here there are ideas, but the correct answer is not yet known.

When the sheet is stretched, the symmetries of the square and circle are not completely broken: The resulting rectangle and oval are both symmetric with respect to rotations about their centers by an angle of 180 degrees. Similarly, not all of the electroweak symmetries are broken—the local electromagnetic symmetry discussed in the last section is unbroken. An important consequence of an exact local symmetry is that it requires the mass of the corresponding force particle to vanish. This explains why gluons and photons are massless. On the other hand, W and Z particles, which correspond to the broken parts of the electroweak symmetry, are not constrained to be massless. In fact, they are so heavy that only in the 1980s did accelerators attain sufficient energies to produce them.

Higgs Boson

The origin of electroweak symmetry breaking, which leads to masses for *W* and *Z* particles as well as for quarks and leptons, is a crucial problem of particle physics. What is doing the stretching? The stretching must be generated by some new interactions of the theory—the known interactions illustrated in Figures 3.4, 3.5, and 3.6 are not able to do the job. In the Standard Model, a hypothetical particle, called the Higgs boson, is introduced and given interactions, which allow the elementary particles to become heavy. The Higgs boson is quite unlike either a matter or a force particle. When physicists say that the Standard Model has been verified in thousands of experiments, they are referring to all the processes that result from the force particle vertices of Figures 3.4 through 3.6. The Higgs boson is still a matter of speculation, lacking solid experimental support. Nevertheless, something must generate particle masses, and physicists know that this physics is inextricably linked to the mass scale of the *W* and *Z* particles.

Generation-Changing Interactions

The interactions that generate the quark and lepton masses play a role in a small but very significant property of the weak force. This is an interaction that causes the generation of a particle to change, as illustrated in Figure 3.7. The

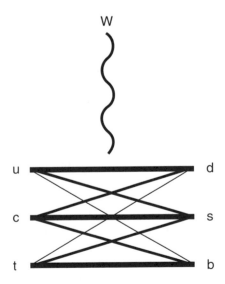

FIGURE 3.7 Generation structure of the weak force.

TABLE 3.2   The 18 Parameters of the Standard Model

Quantities	Number of Parameters
Fundamental electroweak mass scale	1
Strengths of the three forces	3
Masses of electron, muon, and tau	3
Masses of u, c, and t quark	3
Masses of d, s, and b quark	3
Strengths of flavor-changing weak force	3
Magnitude of CP symmetry breaking	1
Higgs boson mass	1

dark lines represent the strongest part of the weak force, which does not change generation. The regular and faint lines represent smaller pieces of the weak force, which are called flavor-changing interactions and are described by three parameters. Experiments have not uncovered any flavor-changing interactions of leptons.

Are the laws of physics invariant under the interchange of particles and antiparticles? If so, there would be a new symmetry of nature, known as CP. The masses and interactions of the particles are nearly identical to those of the antiparticles, but there is a small difference—CP is not an exact symmetry of nature. Breaking of the CP symmetry has been observed as a very small difference in neutral K meson decay probabilities. Within the Standard Model, it is the interactions of the Higgs boson that break CP symmetry, an origin for CP breaking that must be viewed as speculative. This breaking is described by a single extra parameter that enters the flavor-changing vertices of weak interactions. The parameter is capable of describing all of the CP violation observed to date. New experiments studying K and B mesons will soon test whether the generation-changing parts of the weak interaction, illustrated in Figure 3.7, really do provide the answer to the puzzle of the origin of CP violation.

A good theory allows calculations that predict many phenomena in terms of just a few free parameters, which must be measured. The Standard Model has been used to calculate thousands of phenomena in terms of the 18 independent parameters listed in Table 3.2. These are the few quantities that cannot be calculated within the Standard Model.

There is a limit to the accuracy of predictions resulting from calculations in the Standard Model. Frequently this is just because high-accuracy calculations are lengthy. In these cases, great effort can produce extraordinarily precise predictions. For example, the motion of electrons in magnetic fields has been successfully predicted to one part in a trillion. (The measurement is also a great effort!) Calculations of processes induced by the weak force have been computed to better than one part in a thousand and have been verified at experiments carried out at Fermilab, CERN (the European Laboratory for Particle Physics), and the Stanford Linear Accelerator Center (SLAC) over the past 5 years.

For the strong force, calculations are more difficult. Over the last decade, the rapid increase in the speed of computers has allowed remarkable progress; for example, the masses of the proton and rho meson have been computed to an accuracy of about 10%. As yet, these calculations are far from yielding a quantitative understanding of more complex phenomena such as the detailed structure of the proton, but many important calculations are under way.

The Standard Model represents an astonishing synthesis of our understanding of the properties and interactions of elementary particles. The next two sections describe how physicists, inspired by its success, are attempting to understand fundamental laws at a deeper level, with greater conviction than ever before that new symmetries remain to be discovered.

BEYOND THE STANDARD MODEL

Two questions are paramount in furthering an understanding of particles and their interactions, and both of these involve the masses of the particles.

The first question involves the masses of force carriers. The massless photon can be understood in terms of the electromagnetic symmetry, and the mass of the proton follows from the dynamics generated by the strong symmetry. Hence, it is only the mass scale of weak interactions, which leads to heavy W and Z particles, that is not constrained by a symmetry principle. Without such a symmetry, it is not just that the mass scale of weak interactions cannot be determined by theory; rather, the theory naturally makes the mass scale huge, many orders of magnitude larger than observed in nature. The theory can be made to agree with observation only if several large contributions to the weak mass scale are made to cancel, which is an unnatural fine-tuning.

The second question involves the pattern of masses and interactions of the matter particles, shown in Tables 2.1 and 3.1, together with the 18 parameters indicated in Table 3.2. What determines this structure and the values of these parameters? Could a larger symmetry be responsible for grouping the particle in generations, and could such a symmetry provide an understanding of the pattern of interaction strengths and particle masses?

Symmetry Breaking and Supersymmetry

The first question, which is about how symmetries break, is considered now. Physicists are sure that there are new forces responsible for symmetry breaking, and these new forces should themselves be governed by a new symmetry. One possibility is an extension of space-time symmetry, known as supersymmetry. A second is another local internal symmetry, which physicists call technicolor sym-

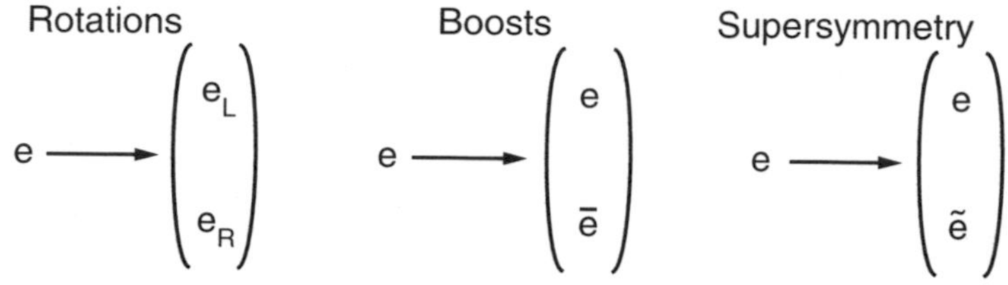

Rotations   Boosts   Supersymmetry

$$e \longrightarrow \begin{pmatrix} e_L \\ \\ e_R \end{pmatrix} \qquad e \longrightarrow \begin{pmatrix} e \\ \\ \bar{e} \end{pmatrix} \qquad e \longrightarrow \begin{pmatrix} e \\ \\ \tilde{e} \end{pmatrix}$$

FIGURE 3.8 Space-time properties of the electron. Rotation symmetry leads to electrons with both left- and right-handed spin. Symmetry under velocity changes leads to a further doubling of the particles, with the electron partnered with its antiparticle. Supersymmetry would lead to still one more doubling: The electron would be partnered with its superpartner.

metry in analogy with the strong color symmetry. Another possibility is that it could be some new scheme that has yet to be invented.

Supersymmetry adds new dimensions to space-time with coordinates that are not ordinary numbers but have a quantum mechanical character. The breaking of such a symmetry could provide an origin for the weak scale. As indicated, as space-time symmetries get larger, the number of states associated with a particle, such as an electron, also increases. It is therefore no surprise that the further extension of space-time symmetries to include supersymmetry leads to a further doubling of the kinds of particles, as illustrated in Figure 3.8. For example, the electron has a superpartner, called the selectron. Moreover, Higgs particles are required.

Technicolor, if it exists, would be a new strong force—similar in many ways to the known strong, or color, force. In the same way that the strong force is responsible for the masses of the proton and other hadrons, so the strong technicolor force could provide masses for the W and Z particles.

As elementary-particle physicists look beyond the Standard Model, they expect to discover a new force. The symmetries for the new forces differ greatly in their predictions: For example, supersymmetry incorporates the Higgs particle of the Standard Model as the origin for quark and lepton masses, whereas in technicolor theories there is no Higgs particle. Theoretical difficulties in constructing complete technicolor theories of nature have led many physicists to see supersymmetry as the most likely option. If supersymmetry does provide the key to the weak scale, then the early decades of the twenty-first century will be a time of great discoveries for particle physics: many new particles, the superpartners of particles, and observations of many new effects in rare processes. The most exciting prospect is that measurements of the masses and interactions of the new superpartner particles will shed light on another great puzzle—the pattern of quark and lepton masses. If technicolor forces are discovered, the future will be even more interesting. As well as a whole new hadron spectroscopy, additional new forces of nature must be present to generate masses for quarks and leptons. As experiments reach toward the answer to the great question of how the weak symmetry is broken, physicists anticipate the possibility of dramatic developments in the future direction of the field.

Grand Unification

The second question introduced at the beginning of this section concerns the origin of the multiplicity of particles, forces, and masses. Progress can be made by a conceptually straightforward extension of the use of local internal symmetries.

If a generation is considered in more detail, including the colors of the quarks, one finds that it has 15 particle components. The three local symmetries of the Standard Model distinguish between these components: Some feel strong and weak forces, whereas others do not, so it is natural to arrange these components into five groups (see Table 3.1). Is it possible that the local symmetries of the Standard Model are just fragments of a much larger, grand unified symmetry? Remarkably, there is such a symmetry that treats all 15 particles of a generation as components of a single fundamental object. The most remarkable aspect is that the properties of this symmetry lead precisely to each and every number in Table 3.1. Grand unified symmetries provide an understanding of the patterns of particles.

If there is a single large local symmetry treating all members of a generation in an equivalent symmetrical way, why does one not observe a single force acting identically on equal-mass particles u, d, e, and v_e? The grand unified symmetry must break at an energy scale that is larger than has been probed by accelerators. In the same way that the electromagnetic force is the low-energy relic from the breaking of electroweak forces, so the three forces of the Standard Model could be the low-energy remnant of a force based on a larger broken symmetry at higher energies. However, as physicists try to understand nature by introducing larger symmetries, the issue of how these symmetries are broken becomes even more important.

In gauge theories, the force between two particles, governed by the interaction strength g, depends slightly on the energy at which the particles collide. At very high energies the grand unified symmetry is unbroken and relates the three interaction strengths, in fact making them equal, $g_1 = g_2 = g_3$, as the three forces become a single force. However, at lower energies, where today's experiments are performed, the grand unified symmetry is broken and the three interaction strengths have different dependencies on particle energy, as shown in Figure 3.9. A combination of g_1 and g_2 (measured at the energy scale of weak interactions) is predicted to be in the range 0.230 to 0.236 if the theory is supersymmetric and in the range 0.212 to 0.218 if it is not. Thus, grand unified theories can precisely predict this quantity, which in the Standard Model could take any value in the range 0 to 1. It has been measured at CERN and SLAC to be 0.231, providing

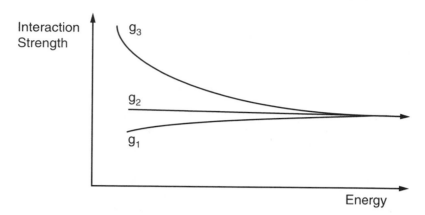

FIGURE 3.9 Unification of strong, weak, and electromagnetic forces. The strengths of the three forces g_1, g_2, and g_3 depend on the energy at which measurements are made. This dependence has been observed experimentally and can be calculated theoretically. Values for g_1, g_2, and g_3, measured at the energy scale of weak interactions, can be extrapolated theoretically to high energies where, if the theory is supersymmetric, they are found to meet, providing a visual picture of the unification of the three forces.

evidence for both grand unification and supersymmetry. Perhaps the most dramatic prediction of grand unification is that protons—a fundamental building block of all matter—are not stable, but decay into lighter particles. The simplest nonsupersymmetric theories have been excluded by experiments that searched for, but did not find, proton decay. The supersymmetric theory predicts a longer life for the proton—only a few in a hundred thousand tons of matter (equivalent to a large battleship) will decay each year. Other phenomena could also probe the structure of these supersymmetric grand unified theories: Neutrinos may have mass, and muons may be converted to electrons when they are close to an atomic nucleus.

Why Are There Three Generations?

The muon and tau are identical to the electron, except that they are much heavier. Why should these heavy copies of the electron exist? Why are there three generations of matter as shown in Table 2.1? Physicists once again look to symmetries for the answer. Consider an equilateral triangle drawn on an elastic sheet. This is an analogy for a symmetric world in which there are three identical charged leptons, with one side of the triangle representing each lepton. The symmetry of the triangle under a rotation of 120 degrees shows that the three leptons are identical. If the triangle is stretched, the three sides are no longer equal, analogous to our world where the electron, muon, and tau have very different masses. As discussed earlier, physicists believe that a symmetrical theory lies behind nature, but that this symmetry is broken by the stretching that produces the familiar world in which we live. The symmetry is not manifest directly, but its presence is inferred from the stretched forms that have broken the symmetry. To understand why nature looks the way we find it, we have to understand broken symmetries. Even if we are able to uncover these broken symmetries of nature, physicists ask why the symmetry was there in the first place: Why start with an equilateral triangle? Why not a square or something else? The next section describes work of the past decade that has grappled with such questions.

PHYSICS OF THE PLANCK SCALE

The energy scale at which the strengths of the three forces are predicted (by assuming supersymmetry) to become equal, as shown in Figure 3.9, is very high, about 10^{14} times larger than the energy scale of weak interactions. Above this energy scale, the symmetry becomes so encompassing that quarks and leptons become unified; this is called grand unification. This energy scale is intriguingly close to the famous Planck energy scale, which is about a hundred times larger. Unfortunately, this energy is so high that it can be reached only by theoretical speculation.

The Planck scale of energy is where the gravitational force becomes strong. At low energies, gravity is an incredibly weak force, noticeable only in the influence of large objects such as planets and stars. The gravitational force depends on a particle's mass when at rest, but on its energy when in motion, so that it increases with increasing energy and at the Planck scale is competitive with the other forces of nature. The Planck scale is determined by the three fundamental units of nature: the maximum speed (that of light, c), the quantum of action (Planck's constant, h), and the gravitational coupling constant (Newton's gravitational constant, G). A mathematical combination of these three constants to yield a term whose units are those of energy results in the Planck energy scale. With other combinations, a Planck length, a Planck time, and a Planck mass can be constructed—any fundamental physical quantity that is not described by a pure dimensionless number such as 2 or π. Given the fundamental nature of these constants it is reasonable to suppose that the Planck scale is the fundamental scale of physics. It is remarkable that these quantities, the Planck length of 10^{-33} cm, the Planck time of 10^{-43} s, and the Planck mass of 10^{19} proton masses, are so disparate, not only from macroscopic quantities but also from the basic scale of the Standard Model. Indeed there is a disparity of 17 orders of magnitude between these scales. This disparity gives rise to the question: Why is the weak scale, which determines the overall scale for the masses of observed particles, 10^{15} times smaller than the Planck scale? Not only is a new symmetry needed to govern the weak scale, but the question arises as to why the energy associated with this new symmetry is so low.

If the purported unification scale is so close to the Planck length, then gravity must be treated on an equal footing with the other forces of nature. This is a long-outstanding problem of theoretical physics. Einstein's theory of gravity is remarkably successful at low energies, yet it gives rise to deep problems and inconsistencies at high energy. These problems suggest that it must be replaced by a more fundamental theory, reinforcing the view that new physics will appear close to the Planck length that might unify all the forces of nature, including gravity. Fortunately, there exists a theory that appears to have the potential of achieving these goals—string theory.

String Theory

What is string theory? String theory says that if we could look at a quark with a microscope that can resolve distances of 10^{-33} cm, we would not see smaller subobjects, but rather a quark would look to us like a little closed string.

String theory is a natural generalization of previous theories of particles but represents a radical departure from the tradition initiated by Thales of Miletus. In uncovering string theory about 25 years ago, physicists set out on a path whose end we can still barely conceive, one that has led to a trail of theoretical surprises—including supersymmetry—without obvious historical parallel.

String theory not only eliminates the contradiction between gravity and quantum mechanics but in a sense explains why just this combination exists in nature. String theory also automatically generates all of the ingredients that seem to be needed as building blocks of the Standard Model. In these and other ways, string theory provides potential answers to many of the puzzles posed by the Standard Model.

Two major revolutions in physics have already occurred in this century: relativity and quantum mechanics. These were associated with two of the three really basic parameters of physics: the velocity of light and Planck's quantum of action. Both revolutions involved major conceptual changes in the framework of physical thought. In each case, the new theory was totally different from the old in its basic tools and concepts, but it reduced approximately to the old one when the appropriate parameter could be considered small.

The last parameter of this sort is Newton's gravitational constant. A third revolution appears to be likely, and string theory—which reduces to more familiar theories at large distances—may be the key. Perhaps this third revolution will lead to a final theory or perhaps only to a next theory that will lead to new questions.

The present state of theoretical physics is reminiscent of the days of confusion that preceded the birth of quantum theory in the mid-1920s, when it was clear that a new theory was coming but not at all clear what this theory was. In the present case, a whole host of theoretical insights clearly point toward a basic change in all of the concepts of space and time. One should not underestimate the likely scope of this change.

String theory is now in the midst of intense theoretical development. Although it appears to have the potential of reproducing the Standard Model and explaining its structure and parameters, the understanding is too primitive to be able to make complete predictions about details of the Standard Model; however, the main qualitative properties of the Standard Model have been derived from string theory in a strikingly elegant way. Moreover, string theory requires the existence of both quantum mechanics and gravity, whereas previous theories in physics make it impossible to have both together; other general predictions of string theory are gauge invariance, which has been seen to be the bread and butter of the Standard Model, and supersymmetry, which is one of the main targets in the worldwide enterprise of particle physics. Many deep problems remain to be solved before the theory can be compared directly with experiment. Nonetheless string theory is testable by experiment. It would be easy for new experimental discoveries that did not fit into a straightforward extrapolation of the Standard Model to provide evidence that string theory is the wrong theory to follow. Conversely, the discovery of supersymmetry would be an important validation for string theory. In addition, this discovery would provide invaluable clues as to the mechanism of supersymmetry breaking that could help in unraveling the predictions of string theory.

Thus, we have the beginnings of a new theory of fundamental physics— string theory—whose full elucidation could be as revolutionary as the discovery of quantum mechanics or relativity.

4

❖

The Past 25 Years:
Establishing the Standard Model

INTRODUCTION

This chapter summarizes the most important experimental results of the past 25 years, those that established the Standard Model as the correct description of essentially all phenomena in the field of elementary-particle physics. Some experiments were designed to test predictions of the new theoretical framework; other discoveries came as surprises. All of the results depended on the construction of new accelerators and the development of new experimental techniques to exploit these accelerators.

THE WORLD OF ELEMENTARY-PARTICLE PHYSICS CIRCA 1972

In 1972, both theoretical and experimental particle physics were unknowingly on the verge of vast and exciting changes. A crude but effective model existed to explain how hadrons were held together by strong interactions, and there were rules to allow the calculation of weak interaction processes. Yet quantum electrodynamics (QED) was the only example of a precise theory that could explain a wide range of experimental results. Papers proposing the electroweak theory were starting to gain wide attention, however, and they were the first harbingers of changes to come.

Of the particles now recognized as truly fundamental that are shown in Tables 2.1 and 2.3, only the first two leptons, their accompanying neutrinos, three quarks, and the photon had been observed by 1972. Even the idea that protons and neutrons were made of quarks was still relatively new: The experiment providing the first evidence for this occurred in 1969.

Although physicists did not realize it at the time, the field of particle physics was about to undergo a revolutionary change caused by the combination of new accelerators and the new theoretical tools called gauge theories. In 1972, the proton accelerator at Brookhaven was delivering beams of unprecedented intensity. The proton accelerator at CERN (the European Laboratory for Particle Physics) was being used to produce intense beams of neutrinos. A new high-energy proton accelerator at Fermilab was just beginning to produce new physics results, and SPEAR, an electron-positron collider, was starting to operate at the Stanford Linear Accelerator Center (SLAC).

THE FORCES

There have been two major developments in the understanding of fundamental forces over the past 25 years. One was the establishment of the idea that electromagnetic and weak forces are unified into a single force that could be described by a theory formulated around the principle of gauge invariance: a gauge theory. The other was the discovery of a theory of the strong force, the force that holds quarks together in the proton, that was similarly a gauge theory.

The Electroweak Force

A central idea of the Standard Model is that the electromagnetic force and the weak force are different manifestations of a single unified force called electroweak. This was not evident for many years because the weak force acts over only very short distances and is completely negligible at the atomic distance scales at which the electromagnetic force acts to bind electrons to the nucleus. Electroweak theory, developed in the 1960s, gave a natural explanation for this difference. The reason for the difference is that the force carriers for the weak force, W and Z bosons, are extremely massive, whereas the force carrier for the electromagnetic force, the photon, is massless. This theory predicted many new phenomena that could be explored with the new, higher-energy accelerators whose use was beginning in the early 1970s.

Demonstrating the Unification of Weak and Electromagnetic Forces

A dramatic prediction of electroweak theory was that there would be a new kind of interaction involving quarks and leptons, called the neutral weak current. One manifestation of this new interaction was that a neutrino could strike a quark or an electron and recoil, remaining a neutrino. Up to this point, the only neutrino interactions observed were those in which a neutrino was transformed

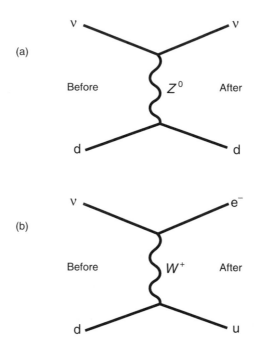

FIGURE 4.1 Two interactions of neutrinos with quarks. In (a), the neutrino scatters from a *d* quark, remaining unchanged. This is an example of the weak neutral current. In (b), the neutrino changes to an electron, and the *d* quark changes to a *u* quark. This is an example of the more common weak charged current.

into a charged lepton, like the electron. These two types of scattering are illustrated in Figure 4.1.

Experimentally, observing an elastic scattering of a neutrino is extremely difficult because the neutrino leaves no trace. The scattering can be observed only by seeing that a proton sitting in a neutrino beam is given a kick, without any charged lepton emerging from the event. In 1973, an experiment at CERN produced the first evidence of neutral current scattering, which was soon confirmed by experiments at Fermilab. The number of events observed was consistent with the prediction of the electroweak theory, and this discovery provided the first real evidence that theorists were on the right path. Theorists had successfully predicted the existence of new particles, but this was the first time that a fundamental particle physics interaction had first been predicted by theory and then discovered experimentally.

From that time on, a wide range of many different types of experiments studied neutral currents, and electroweak theory was able to accurately predict the results of all of them. Some experiments used neutrinos, and others were able to measure the effect of neutral currents in scattering electrons off quarks. A precise measurement of electron scattering at SLAC saw a tiny difference of about one part in one hundred thousand in the scattering rate of left- and right-handed electrons, exactly as predicted by electroweak theory. Evidence that the electromagnetic and weak forces were unified was overwhelming; the next task was to actually discover the predicted gauge bosons of the weak interaction.

Discovery of W and Z Bosons

The second major prediction of electroweak theory was the existence of W and Z bosons. Results from neutrino experiments could be used to predict their masses to be around 80-90 GeV (1 GeV = 10^9 electron volts), and the search for these bosons was given the highest priority. However, a new breakthrough in accelerator technology was necessary. The experiments of the 1970s, which used proton beams hitting stationary targets, could not produce a particle with a mass greater than about 10 GeV. To reach higher energies, it was necessary to use a proton-antiproton collider, at which much higher energies were accessible.

Starting in the mid-1970s, the Super Proton Synchrotron (SPS) accelerator at CERN was converted to a proton-antiproton collider capable of reaching an energy of 540 GeV in the center of mass of the collision, about 20 times the energy possible in fixed-target experiments. For the first time since the Bevatron was built in the 1950s to produce the antiproton, a new accelerator was built with the express purpose of discovering a new particle predicted by theory. Two experiments were constructed to observe the rare events in which bosons were produced and then decayed to leptons. In 1983 W and Z bosons were both discovered at CERN, with masses in the range expected from electroweak theory. Decades after the discovery that the photon had no mass, its massive siblings— the gauge bosons of the weak force—were observed in the laboratory.

Precision Tests of the Electroweak Force

During the past decade, two new electron-positron colliders dramatically improved our understanding of weak interactions. An enormous collider at CERN (called the Large Electron-Positron collider [LEP]) of the conventional circular type produced millions of Z bosons. A much smaller linear electron-positron collider at SLAC (called the Stanford Linear Collider [SLC]) produced fewer Z bosons, but its innovative design allowed experiments with polarized beams, where the spins of beam particles were aligned to a common orientation (SLC was also important as a prototype for a possible future linear collider, discussed in Chapter 6). For the first time, precise measurements of the fundamental parameters of electroweak theory could be made. These measurements could then be used to probe its validity, in much the same way that precise tests of electromagnetic theory have been made for 50 years. One outcome of these studies was that LEP and SLC experimentalists were able to put upper limits on the mass of the *t* quark, a sixth quark that had not been observed directly. If the electroweak theory was assumed to be correct, then the relationships between the electroweak parameters measured at LEP and SLC depended on the *t* quark mass. In this way, the *t* quark was indirectly measured to have a mass of 180 ± 15 GeV before it was even discovered. The excellent agreement between this indirect measurement of *t* quark mass and the direct measurement made once it was discovered, provided a stringent test of physicists' understanding of the electroweak force. The current status is that in dozens of measurements to precisions of fractions of a percent, electroweak theory and experimental measurements are in spectacular agreement everywhere.

The Strong Force

The second innovation in the description of forces in the Standard Model is the theory of the strong force, known as quantum chromodynamics (QCD). In 1972, physicists could explain hadrons, such as the proton, as composites of quarks. From detailed experiments at SLAC, it was known that if high-energy electrons were fired into a proton, they would scatter off its quarks, which acted like hard objects much smaller than the proton itself. There was no theory to explain how three quarks would bind together to make a proton or neutron or to explain why isolated quarks were never observed.

Around 1973, however, QCD was developed. It accounted for the observation that quarks are effectively confined inside the proton. The massless force carriers, analogous to the photon, were called gluons, because they provided the "glue" that held the proton together (gluons, like quarks, could not be observed in isolation).

Experimental study of QCD as the theory of the strong force has been much more difficult than studying electromagnetic or weak forces. The traditional calculational tools developed for QED and electroweak interactions are often inadequate for QCD. Nevertheless, QCD has been verified experimentally, and there is continuing progress in developing the calculational tools necessary for precision tests of the theory.

QCD predicted that the strength of the strong force would decrease slightly with increasing energy. In the mid-1970s, this was beautifully confirmed by a series of precision experiments at Fermilab, SLAC, and CERN, which scattered electrons, muons, and neutrinos from protons.

Discovery of the Gluon

An important verification of the theory came from indirect observation of the gluon. The gluon, like quarks, manifests itself in high-energy collisions as a collimated jet of particles. The perfect environment for observing jets is in electron-positron annihilations at high energies. After the spectacular success of SPEAR in the early 1970s (see below), electron-positron colliders with about 35 GeV energy were built at the German laboratory DESY and at SLAC. Most of the events were of the type shown in Figure 4.2(a). Two quarks are produced in the collision that form two jets of particles with equal energy and opposite directions. QCD predicted that at high-energy colliders, three-jet events from the process illustrated in Figure 4.2(b) should also be observed; the third jet comes from a gluon produced at high energy. Figure 4.3 shows an example of such a three-jet event, collected in 1978 at DESY. This was dramatic confirmation of a prediction of QCD.

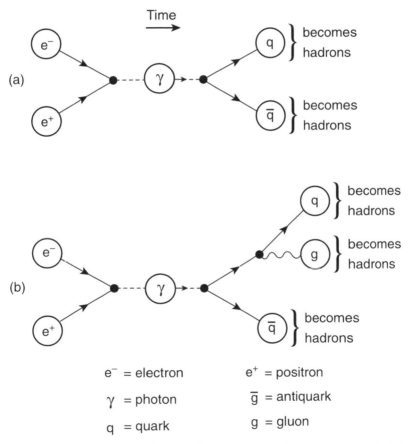

FIGURE 4.2 Two examples of how an electron and a positron annihilate. In (a), an electron-positron pair is annihilated to a quark-antiquark pair. In (b), one of these quarks also emits a gluon.

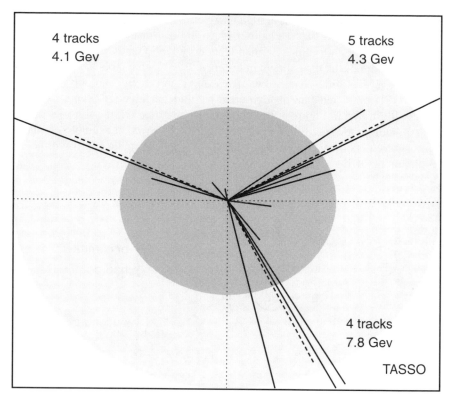

FIGURE 4.3 Event showing the three-jet structure used to discover the gluon predicted by quantum chromodynamics. (Courtesy of Sau Lan Wu, Two-Arm Spectrometer Solenoid [TASSO] Collaboration.)

Strength of the Strong Interaction

A fundamental property of all forces is their strength. The strength of the strong force has been very difficult to measure, however, in part because quarks and gluons are confined. Nevertheless, a series of measurements of the strength of the strong interaction (g_3) was made, starting in 1978, using very different techniques. It is a severe test of the theory that all of these measurements agree and determine this fundamental constant of nature rather well. Also, it has been shown that accurate measurements of the strength of the forces can be used to test ideas of "grand unification," the possibility that all of the forces derive from a single one. As calculational tools mature, these measurements are continuing.

The Spectrum of Particles

High-energy physicists use the masses of hadrons to probe the strong inter-action that binds quarks to form particles in the same way that atomic physicists early in this century used the spectra of atoms to study the electromagnetic interaction that binds electrons in the atom. As a result of a concerted experi-mental effort over the past 25 years, approximately 50 new quark bound states have been discovered at a variety of high-energy machines around the world. QCD has the potential to calculate the particle spectrum in terms of the quark constituents and fundamental equations describing the behavior of the strong interaction. This is a difficult and challenging task. One continuing mystery is that QCD predicts an even richer spectrum of states than has so far been ob-served. Quantitatively explaining the wealth of experimental data from QCD will continue to be a challenge for theorists of the next decade.

CONSTITUENT PARTICLES

Progress over the past 25 years in understanding the constituent particles has been just as dramatic as for the fundamental forces. In 1972, only the first two generations of leptons had been observed: the electron, muon, electron neutrino, and muon neutrino. Three types of quarks were known: up, down, and strange. Experiments since that time have discovered the tau lepton and its neutrino, as well as three more quarks: charm, bottom, and top. In addition, experimental studies of Z-boson decay have determined that there are exactly three neutrinos, and therefore the familiar pattern of generations of quarks and leptons shown in Table 2.1 will *not* continue any further.

Discovery of the Charm Quark

In 1972, knowledge about quarks was still rudimentary. The breathtaking discovery of the charm quark in 1974 was hailed as the beginning of "the new physics." The experimental evidence was dramatic, and in an astounding quirk of fate, the charm quark was discovered simultaneously at two different labora-tories with very different experiments. The Mark I collaboration at SLAC ob-served the charm quark using the electron-positron collider SPEAR. At the same time, a group at Brookhaven using protons on a fixed target in a com-pletely different type of experiment also detected the new quark. The discovery established that the structure of repeating generations seen in the leptons also applied to quarks.

Discovery of the Tau Lepton

When the muon was discovered in 1936, physicists wondered why nature created this heavier copy of the electron. It was the first indication of the repeat-

FIGURE 4.4 The Mark I experiment, which operated at the electron-positron collider SPEAR at SLAC from 1972 to 1976. Many groundbreaking discoveries were made using this instrument, including discovery of the tau lepton and codiscovery of the charm quark. It served as the prototype for the next generation of experiments. (Courtesy of the Stanford Linear Accelerator Center.)

ing generations of quarks and leptons. In 1976, still flush from the discovery of the charm quark, physicists on the Mark I detector operating at SPEAR discovered a third-generation lepton, the tau. Figure 4.4 shows the Mark I detector at SLAC in which this and many other discoveries were made. After years of subsequent research, the properties of the tau lepton have been measured to be precisely as predicted for a heavier repetition of the electron, leading to the conclusion that the three generations of charged leptons are distinguished only by the large differences in mass.

Discovery of the Bottom Quark

The charm quark completed the second quark generation. As soon as evidence for a third generation of leptons was found with the discovery of the tau lepton, physicists intensified their search for more quarks. However, even if such a third generation of quarks did exist, there was no guidance from theory as to what the mass of third-generation quarks might be.

The bottom quark was discovered in 1977 in an experiment using the still-new proton accelerator at Fermilab. Figure 4.5 shows the experimental signal of the discovery. Thus, two new quarks and one new lepton were discovered at American accelerator laboratories in a period of less than 3 years, after a period of 25 years without any new particles of this type.

Starting in 1981, the energy of the electron-positron collider CESR (the Cornell Electron-positron Storage Ring) at Cornell University was tuned to a value that maximized the number of *B* mesons, particles containing a *b* quark, that could be observed. The steadily increasing collision rate of this collider, along with a similar one called DORIS at the German laboratory DESY, made possible ever more precise measurements of the decays of the bottom quark. This program has continued until the present, and CESR now has the highest rate of collisions of any collider in the world. Measurement of properties of the bottom quark continues to be a very active part of the current high-energy experimental program.

Experiments showed first that the *B* meson had a lifetime longer than expected and then that there was significant mixing amplitude for *B*, anti-*B* oscillations. These unexpected findings were crucial for the expected viability of performing CP measurements in the *B* meson system and stimulated the design of *B* factories.

With the discovery of the bottom quark, a new question jumped to the top of the list for further experiments to answer: What is the mass of the quark that is the partner of the bottom quark in the third generation?

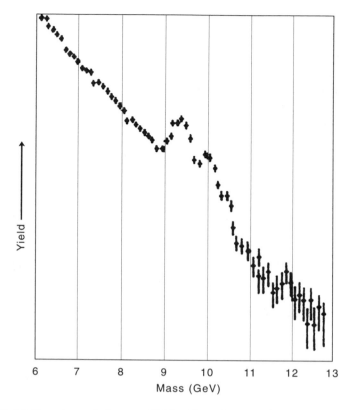

FIGURE 4.5 Experimental data establishing the existence of the bottom quark. The relative frequency of producing muon-antimuon pairs in proton-proton collisions is shown to decrease rapidly with the mass of the muon-antimuon pair. An anomalous peak occurs at a mass of about 9.7 GeV, however, which is due to the production of a particle with this mass made of a bottom quark-antiquark pair; this then decays to a muon-antimuon pair. (Courtesy of Leon Lederman, Illinois Institute of Technology and Fermilab.)

Discovery of the Top Quark

The top quark was by far the most elusive of all. The top quark was a necessary component of the Standard Model of electroweak interactions, but there was no consistent theoretical guidance as to what its mass should be. By 1988, the search had extended to the lofty mass of 41 GeV, almost 10 times the mass of the bottom quark, with no success. With the onset of operations at LEP and SLC in 1989, physicists could begin to extract limits on the top quark mass from very precise measurements of the properties of the Z boson. By 1992, it was indicated that the top quark mass must be between 100 and 200 GeV if the

Standard Model was correct. This was a startlingly high mass, given the masses of other quarks.

The only way to observe a top quark with such a high mass was at the collider with the highest energy, the Tevatron antiproton-proton collider at Fermilab. With each run, the rate of collisions at the Tevatron was increased, allowing an extension of the search for the top quark. Finally, in 1994, the Collider Detector at Fermilab (CDF) experiment announced the first evidence of top quark production. The existence of the top quark was firmly established in 1995 with simultaneous announcements by both the CDF and the D0 experiments of results that demonstrated a mass of around 175 GeV for the top quark. Figure 4.6 (in color well following p. 112) shows a top quark event from the CDF experiment. After an 18-year search, the last quark (at least within the Standard Model) was found.

Counting the Number of Generations

The pattern of generations, each with a charged lepton, a neutrino, and two quarks, has been repeated three times. Could there be more? An example of the important physics coming from the Z factories, accelerator facilities that produce large numbers of Z decays, was the counting of lepton generations. The Z decays very quickly into any lepton or quark and its antiparticle (except for the top quark, which is too heavy). By very precisely measuring the rate at which the Z decays, experiments were able to determine the number of neutrino types and found that there are exactly three. (In principle, there could be more generations, but the associated neutrinos would have a mass of more than 45 GeV.) This means that the pattern of repeating lepton generations, each with a heavier lepton than the one before, stops at three, and since in the Standard Model it is necessary to have the same number of leptons and quarks, all the leptons and quarks that nature has to offer may have been found. Such a conclusion would have been impossible before the framework of the Standard Model had been developed. Figure 4.7 shows some of the data that led to measurement of the number of neutrino types.

PARTICLE-ANTIPARTICLE ASYMMETRY

In 1964, physicists made an unexpected and astounding discovery: They found a difference in the ways that matter and antimatter behave. They observed about 45 K meson decay events that would have been forbidden if particles and antiparticles behaved symmetrically. Until this astonishing experiment, it had always been assumed that the laws of nature operate identically on matter and antimatter.

This particle-antiparticle asymmetry known as CP violation is one of the most profound mysteries of particle physics. There are only two manifestations

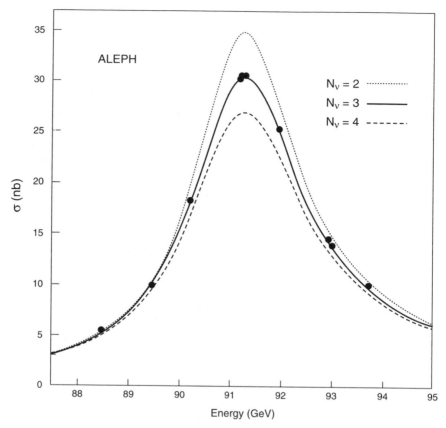

FIGURE 4.7 Probability of an electron and positron annihilating into quarks, plotted as a function of the energy, near 91 GeV. The peak occurs at the mass of the Z boson. The width of the peak gives a precise measurement of the rate at which Z decays, which in turn specifies the number of neutrino types. (Courtesy of the Apparatus for LEP Physics [ALEPH] collaboration, CERN.)

of this violation known to date. One is the CP violation discovered in the decay of K mesons in 1964, which has been under continual investigation since. The other is the fact that there is matter left over from the "big bang," which is in the form of stars and planets. If there were a perfect symmetry between matter and antimatter, equal amounts of each would have been produced in the very early universe and would have largely been annihilated, leaving little raw material on which to build structure in the universe.

Description of the transitions of quarks from one generation to another provided a framework for understanding CP violation. In this model, CP violation could be described by the same parameters that describe quark transitions from one generation to another, as discussed in Chapter 3. Ever more demanding experiments on the *K* meson system have explored the asymmetry in greater detail throughout the past 25 years and continue through the present. These experiments are all consistent with the source of CP violation being in the flavor-changing quark transitions, although some other explanations still survive. Furthermore, the search for an understanding of particle-antiparticle asymmetry has motivated many detailed studies of the bottom quark in an experimental program that started in the early 1980s and continues to be very active today.

OTHER STUDIES

Measuring the Mass of Neutrinos

It has long been realized that neutrinos are peculiar. All of the other leptons and all of the quarks are massive. Neutrinos, on the other hand, have masses that are either zero or so small as to be, thus far, experimentally indistinguishable from zero. However, there are no known reasons for the neutrino mass to be exactly zero, so for the past 25 years, there has been a very dedicated experimental effort to look for a mass for any of the neutrinos. This effort continues today.

Direct measurements of neutrino masses are very difficult, and many different techniques have been employed. The best limits are on the mass of the electron neutrino, coming from studies of the spectrum of electrons in tritium beta decay. In a stroke of good fortune, the spectacular explosion of the supernova SN1987a also helped to limit the electron neutrino mass. The supernova released a burst of neutrinos that were detected in large experiments deep underground that had been designed to search for proton decay. From the distribution of arrival times of neutrinos from SN1987a, a limit on their mass could be deduced. Today, the electron neutrino mass is known to be less than 0.003% of the electron mass itself.

However, if the neutrino masses are different from zero even by such a small amount, then they can undergo transitions from one generation to another just as quarks do. For example, if a neutrino is produced as a muon neutrino, it could change into an electron neutrino with time. This process is called "neutrino oscillations," and its rate depends, among other things, on the masses of the neutrinos involved. Oscillations give a method of seeing the effect of nonzero neutrino mass at levels well below what could be measured directly.

The strongest evidence that neutrinos may oscillate, and therefore have nonzero mass, has come from solar neutrinos. In a truly groundbreaking experiment located deep in the Homestake mine in South Dakota, physicists have collected the first evidence of neutrinos originating from the Sun. However, the number of neutrino interactions they see is well below the calculated rate of solar neutrinos that should reach Earth. This discrepancy could occur if electron antineutrinos, which form the bulk of solar neutrinos, oscillate to some other neutrino type. New experiments that could search for these neutrinos in other ways began to obtain data in the late 1980s. By now, five experiments using three different techniques have observed a similar deficit in the rate of solar neutrinos reaching Earth. These results are not yet claimed as final evidence of neutrino mass, however. New experiments are about to start that can measure the rate of all neutrino types together. They could provide the last bit of information necessary to nail down the discovery that the solar neutrino deficit comes from neutrino oscillations; this in turn would imply that neutrinos have a mass.

Searching for Proton Decay

During this period, several ingenious experiments were built to look for the very rare decays of protons in matter, which is predicted if all forces are unified. For example, one experiment consisted of an instrumented tank of 8,000 m^3 of purified water in a salt mine near Cleveland. After years of searching, no proton decays were seen, thus eliminating the simplest grand unified theories from contention. In this case, successful experiments shaped particle physics by discovering that something did *not* happen.

Other Physics Beyond the Standard Model

One way of searching for physics at mass scales above the reach of accelerators is to search for rare decays. These decays would be caused by very heavy particles that interact in ways forbidden in the Standard Model. To observe rare decays, very large data samples are needed and such searches often push the limits of available accelerator and detector technology. Such programs have been under way at Brookhaven National Laboratory, Fermi National Accelerator Laboratory (FNAL), the Japanese High-Energy Accelerator Research Organization (KEK), and CERN in searches for very rare K meson decays; at CESR, for rare B meson and tau lepton decays; and at Los Alamos, for rare muon decays. No evidence has been found for decays that violate the predictions of the Standard Model, even though rare K decays have been searched for with a sensitivity of better than one part in 10^{11}. Rare decays of the muon have been searched for with a sensitivity of one part in 10^{12}. Again, such studies are important in limiting the scope of new physics.

SUMMARY

Over the past 25 years, particle physics has undergone a period of spectacular development. This period began with a wealth of interesting phenomena and a patchwork quilt of theoretical ideas, each of which explained some part of the data. All of the available experimental data that have been collected are now well described by a theory called the Standard Model, which has been verified experimentally to great precision in an extraordinarily diverse set of measurements. It has proven to be frustratingly accurate. Every experimental result so far either has agreed with the Standard Model prediction or has turned out to be wrong! Some experiments made startling new discoveries, which helped to develop the model, and others made measurements of unprecedented precision in order to test it.

One measure of the breakthroughs recognized during this 25-year period is the Nobel Prizes that have been awarded for experimental or theoretical work in this field. These are listed here:

- Burton Richter and Samuel Ting for the discovery of the charm quark;
- Martin Perl for the discovery of the tau lepton;
- Carlo Rubbia and Simon Van der Meer for the discovery of W and Z bosons;
- James Cronin and Val Fitch for the discovery of particle-antiparticle asymmetry;
- Sheldon Glashow, Abdus Salam, and Steven Weinberg for development of the unified electroweak theory;
- Jerome Friedman, Henry Kendall, and Richard Taylor for the first observation of quarks inside the proton;
- Frederick Reines for the first observation of the electron neutrino;
- Leon Lederman, Melvin Schwartz, and Jack Steinberger for the experiment establishing that the muon neutrino and the electron neutrino are separate particles; and
- Georges Charpak for the development of particle detectors.

Since 1972, high-energy physics has advanced to the stage at which almost the entire Standard Model has been established. Three lepton generations, all six quarks, and the gauge bosons for strong, electromagnetic, and weak interactions have all been observed. All of the fundamental particles have been seen, except the Higgs boson or whatever takes its place.

5

❖

The Physics of the Next Decade

OVERVIEW

Previous chapters have shown that the great successes of the past quarter of a century in elementary-particle physics are embodied in the Standard Model, which contains both the electroweak interaction and the strong force. The former, a unification of weak and electromagnetic forces, has been successfully tested at the 0.1% level, and all carriers of the interaction have been observed and studied. The strong force is described by a fundamental theory whose predictions have been checked at both high-energy hadron and electron accelerators. Also, what are believed to be all of the elementary constituents of matter—three pairs of quarks and three pairs of leptons—have been observed.

However, it has been shown as well that the Standard Model is very likely an incomplete description of nature. Many particle properties are not predicted by the theory, including the masses of quarks and leptons and the way flavors are mixed by weak interactions. Physicists do not know whether the strong and electroweak interactions are different manifestations of a single force, and the origins of the breaking of electroweak symmetry, either by a Higgs particle or by some other mechanism, have not been discovered.

It is the need to answer these and other essential questions that drives the program of the next decade described in this chapter. The facilities in which this program will be carried out are distributed throughout the world as shown in Figure 5.1 and listed in Tables 6.1 and 6.2. They include an upgrade to the LEP electron-positron collider to run at energies up to 190 GeV; an upgrade to the Tevatron at Fermi National Accelerator Lab (FNAL) to produce higher rates of

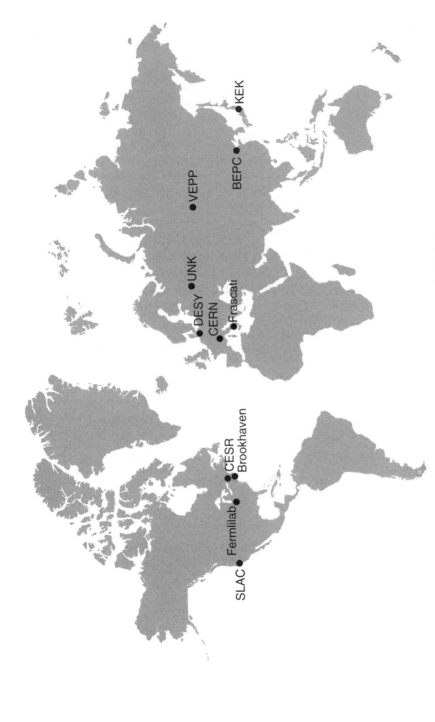

FIGURE 5.1 A map of the world showing the locations of the major high-energy physics facilities.

proton-antiproton collisions at an energy of 2,000 GeV; three electron-positron machines—the Cornell Electron-positron Storage Ring (CESR) at Cornell University, the Positron-Electron Project II (PEP-II) at the Stanford Linear Accelerator Center (SLAC), and the High-Energy Accelerator Research Organization (KEK) *B* factory in Japan, operating at 10.6 GeV to study *B* mesons; and the Large Hadron Collider (LHC), which will come on-line in 2005 and collide protons at energies of 14,000 GeV. The new data will investigate the very underpinnings of the theory. Many questions should be answered by the experimental program of the next 10 years, and there will almost certainly be surprises that alter physicists' view of the world. In Chapter 7, the remaining questions that will have to be addressed at facilities not yet planned are described. The issues addressed here represent only the major thrusts of particle physics today. The list of topics in this chapter is not inclusive, and many other problems are being attacked that are important to a full understanding of the elementary constituents and forces in nature.

WHAT IS THE ORIGIN OF MASS?

As discussed in detail in Chapter 3, the origin of mass is the least understood part of the picture of elementary particles and forces. In the very tightly constructed theories of the strong, electromagnetic, and weak forces, the most natural state would be for all elementary particles discussed to be massless. Indeed, if nature consists of only the presently known particles, then none of the fundamental particles should have mass.

As discussed earlier, one mechanism to describe how particles can have mass requires that at least one more particle exist in nature, the Higgs boson. There could be only one such particle, or there could be several. This might be an elementary particle, or it might be a composite particle in the same way that a proton is made of quarks or an atom is made of electrons and a nucleus. Physicists may find, however, that there is no Higgs particle at all; in this case, some new physics must take its place and give particles mass.

In the simplest case, the Higgs boson is a single elementary particle. The Standard Model does not determine the mass of the Higgs boson, but various lines of reasoning based on empirical data suggest loose upper and lower bounds on its mass. Therefore, experiments now planned or running must incorporate strategies for detecting the Higgs boson within a wide range of possible mass values.

If the Higgs boson has a mass of less than about 95 GeV, it is most easily found in electron-positron annihilation experiments at the LEP collider at CERN (the European Laboratory for Particle Physics). The Tevatron proton-antiproton collider at Fermilab may also be sensitive to Higgs bosons in this range in the next decade. If the Higgs has a higher mass, its detection must await the 14 TeV (10^{12} electron volts) center-of-mass proton-proton collisions in the Large Hadron Collider (LHC) at CERN. Below about 800 GeV, it should be detected directly by ATLAS and CMS, the two large detectors being constructed at LHC. Above this value, it can still be detected indirectly because it will affect the rate for producing pairs of vector bosons. It is a tremendous challenge to design and build an experiment that will discover the Higgs if its mass is anywhere from 95 to more than 1,000 GeV. As the mass changes, the way in which the Higgs will reveal itself changes. Detailed studies have been done to ensure that the LHC experiments are sensitive to the entire range of Higgs mass. The LHC will either observe Higgs particles directly or rule out their existence and provide the first clues to the new physics that is responsible for mass.

WHY ARE THERE ENERGY SCALES THAT ARE SO VASTLY DIFFERENT?

If there is a fundamental theory that explains all the forces that have been observed, the natural energy for describing such a theory is at or near the scale of gravity (10^{16} to 10^{19} GeV). Chapter 3 has already noted that it is extremely difficult to produce a theory that relates physics at the scale of gravity with physics at the 100-GeV scale of the electroweak interaction. This is often called the naturalness or fine-tuning problem because, to achieve a viable theory, it is necessary for a parameter in the theory to have a value that must be specified to 34 significant digits!

Chapter 3 discusses two possible solutions to this problem that have been considered: supersymmetry and technicolor. The more thoroughly explored explanation is supersymmetry (SUSY), which postulates an as-yet unobserved "partner" particle for each elementary particle. If supersymmetry is to solve the naturalness problem, the masses of supersymmetric particles must be less than approximately 1,000 GeV. This puts them in the range of existing or planned accelerators.

Because SUSY contains many new particles, each with a short lifetime and complex decay scheme, there is not a unique way to search for it. On the other hand, many supersymmetric processes produce very distinctive signatures. For example, if a pair of supersymmetric relatives of the carriers of the weak force is produced, three electrons or muons could be seen in a detector after decay and nothing else. The observation of such processes would be strong evidence for supersymmetry since no other known process would create such unusual events.

The search for supersymmetry is being carried out at the Fermilab Tevatron, CERN's LEP collider, and the Stanford Linear Collider (SLC) accelerator at SLAC. Thus far there is little compelling evidence for *any* new processes, including SUSY. This is not surprising since only a small fraction of the possible SUSY mass range is accessible at these accelerators. Between now and 1999, the energy of the LEP collider will increase so that more of the SUSY range will be explored. The increase in beam intensity at Fermilab's Tevatron, beginning in 1999, will further extend the search for supersymmetric quarks and gluons, as well as partners to electrons, muons, and W and Z particles.

When operation begins at CERN's LHC collider in 2005, the available energy in accelerator collisions will increase by a factor of seven, and a huge range of SUSY masses will be accessible to experiment. If supersymmetry is discovered, a great deal will be learned at the LHC. Different types of SUSY particles will be seen, and many of their properties studied. However, it will not be possible to observe all of the SUSY particles and their decays.

If the LHC does not find supersymmetry, it is very unlikely that SUSY is the explanation of the naturalness puzzle. An alternative model to SUSY is a heretofore unobserved very strong force associated with a number of new, high-mass particles; this is the scheme that physicists call technicolor. This model can be distinguished from supersymmetry both by the types of new particles seen and by the way the new particles decay. The LHC and future accelerators will allow a thorough search for a number of types of such particles.

If neither supersymmetry nor a new strong force is observed, then the new high-energy accelerators will reveal the onset of unexpected phenomena. These will serve as signposts to the new physics that produces the enormous difference between the energy scale at which the electromagnetic and weak forces become unified and that above which all forces become unified.

WHAT IS THE ORIGIN OF MATTER-ANTIMATTER ASYMMETRY?

At the moment, studies of K decays offer the only clue to matter-antimatter asymmetry: K mesons are the only experimentally accessible system that has manifested such asymmetry. Physicists believe that if they further study CP violation in the K meson system and in the B meson system, which might exhibit a similar matter-antimatter asymmetry, they can learn something about this cosmological mystery of the universe.

Chapter 3 shows that the Standard Model with three generations might explain why CP violation appears in the K meson system; if this model is correct, similar phenomena should be evident in the B meson system. In fact, the B system should display a rich variety of CP violating asymmetries in many different decay modes, and some of them are predicted to be quite large. For example, the rates for a B meson and its antiparticle to decay to particular final states of interest are very low but may differ by 20% or even more. Furthermore, asymmetries that might be measured in the B system can be related directly to parameters describing quark flavor-changing transitions in the Standard Model, so with enough data, the B system can provide a definitive test of the theory. K and B experiments are in progress or planned at a variety of facilities—Cornell, SLAC, KEK, FNAL, DESY, Brookhaven, Frascati, and CERN—to see if this model is correct.

The experiments that will pursue this study have several common features. The effects are small, so large numbers of particles, either K mesons or B mesons, are necessary. The technology frontier of accelerators is being pushed, particularly in the case of B mesons, to create enough data for these studies. In addition, detectors that study CP violation have to be very sophisticated to reveal evidence for a small matter-antimatter asymmetry in a convincing fashion, and new state-of-the-art detectors are either coming on-line or being built expressly for this purpose. What is the hoped-for outcome of these studies? These experiments should definitively confirm or rule out the current understanding of this phenomenon. Also, if physicists can learn more about the asymmetry between matter and antimatter already evidenced in the K system, and perhaps show that B asymmetry fits into a similar pattern, they may be able to determine the mechanism for CP violation.

In addition to answering the question of where the observed CP violation comes from, physicists are struggling to understand why more CP violation is not observed. This is because there is a natural mechanism to produce CP violation in the strong interactions at a much larger level than that observed.

The symmetries of the Standard Model allow an additional interaction among gluons, which leads to the strong interactions that break CP symmetry. However, the observed breaking of CP symmetry is a tiny effect, so such a strong breaking must be absent. All interactions allowed by symmetries are expected to occur—why should this one be absent? This is known as the strong CP problem. One solution involves a new, hypothetical particle known as the *axion*—new symmetries dictate interactions for the axion, which lead to a cancellation of the CP violating effects in strong interactions.

Axions are predicted to have mass and to interact very weakly with matter; they are believed to have been produced in the very early universe, at the same time hadrons were produced. Searches for them are based on the fact that axions can be detected by their decay into two photons, and a team of physicists from U.S. universities and national laboratories is perfecting an experiment that will reach a significant limit in the next decade.

PATTERNS OF QUARK AND LEPTON MASSES AND TRANSITIONS

Chapter 3 listed 18 input parameters of the Standard Model, which are so far unexplained. There is currently particularly intense interest in measuring all of the parameters of quark flavor-changing interactions, which act to change one type of quark into another (e.g., an s quark to a d quark). The most active work is going on at the CESR accelerator at Cornell; at the soon-to-be-completed PEP-II at SLAC and the KEK B factory; and at the K decay experiments at Fermilab, Brookhaven, and CERN.

If neutrinos have no mass, the Standard Model weak interactions do not change lepton flavor. However, if neutrinos have small masses, then flavor-changing processes are possible for leptons. There are two places in which lepton flavor-changing interactions might be important. First, they could result in leptons that exhibit the same kind of matter-antimatter asymmetry that quarks do. Second, they would allow for the lepton flavor oscillations discussed in Chapter 4, which could then be related to the observed deficit of neutrinos coming from the Sun. Over the next decade, a strong experimental program with several dedicated experiments using a variety of different techniques will search for evidence of neutrino mass and neutrino mixing in order to address the question of whether or not lepton flavors mix.

One technique takes advantage of the Sun as a copious source of neutrinos, with about 60 billion incident on each square centimeter of Earth every second. The effect of the matter in the Sun on the propagating neutrinos makes these searches sensitive to very small quantum mechanical mixing. The Sudbury Neutrino Observatory in Canada and the Superkamiokande experiment in Japan are sensitive to neutrinos from the Sun and will produce measurements that, in a few years, could definitively establish that neutrinos have mass and that lepton flavor-changing interactions occur.

Another approach to this problem uses the cosmic rays that are constantly raining down on Earth. When primary cosmic rays that have traveled great distances through space strike the nuclei of atoms in Earth's upper atmosphere, they produce showers of particles that eventually decay to neutrinos. One can calculate how many of each neutrino type are expected in a detector located deep underground. Two experiments, the Kamiokande and IMB collaborations, observe that the ratio of events containing a muon to events containing an electron indicates that neutrinos change their flavor over the distances involved. The results are statistically significant, and their similarity is striking; however, they have to be confirmed. Progress will come soon from the Superkamiokande experiment, which significantly extends the sensitivity of its predecessor and definitively provides evidence for lepton flavor mixing.

The NuMI (Neutrinos at Main Injector project) experiments being built at FNAL, along with similar efforts in Japan and at CERN, have the potential to definitively resolve the issue of lepton flavor mixing. These experiments are accelerator-based experiments in which beams of neutrinos are produced in a controlled environment. Detectors hundreds of kilometers away from the source then search for direct evidence of neutrinos changing flavor, for example, from a muon neutrino into an electron neutrino or tau neutrino.

The ultimate goal is to explain the patterns of mixing of leptons and quarks, as well as their masses. In addition to precision measurements of the mixing of quarks (and perhaps leptons) in Standard Model weak interactions, it is very important to continue searching for evidence of flavor-changing interactions that do not occur or are highly suppressed in the Standard Model. A variety of dedicated experiments will come into operation over the next decade, which will look for proton decay, muon decay into an electron plus a photon, or CP violation in electrons or neutrons. A positive result from any of these experiments would be fantastically exciting because it would be evidence of physics beyond the Standard Model.

UNDERSTANDING THE STRONG FORCE

Although there is general agreement that the basic elements of the theory of the strong force, quantum chromodynamics (QCD), are correct, major unsolved problems remain. It has not yet been fully demonstrated that quarks are "confined." Already mentioned is the puzzle that QCD naturally accommodates a matter-antimatter asymmetry far beyond what has been observed. And many experimental manifestations of the strong force cannot yet be adequately predicted because of limited calculational tools.

Since QCD is involved wherever there are quarks and gluons, almost all experiments involving these constituents are relevant. Experiments at Fermilab are probing QCD at the very highest energies, where calculations are thought to be most reliable. Experiments at electron-positron colliders provide valuable inputs on the low-energy behavior of QCD. Several experiments, however, have been or soon will be constructed with an emphasis on exploring QCD, and there already have been some surprises.

The HERA facility, at the DESY laboratory in Hamburg, Germany, is the world's first and only electron-proton collider. HERA is the next giant step in a fruitful line of experiments in which leptons (electrons, muons, neutrinos) are scattered from protons. These experiments measured how quarks and gluons are configured inside the proton, a property that cannot yet be calculated accurately from basic QCD principles. By using the technology of modern colliders, HERA has taken lepton-nucleon scattering into new regimes, and new things are being learned about how the proton is constructed.

In addition to the three component quarks and the gluons that bind them in a proton, additional quarks and gluons can exist as brief "virtual" particles, whose existence can be detected in an energetic collision. Experiments have observed a remarkable increase in the numbers of these quarks and gluons that carry only a very tiny fraction of the proton energy. This trend cannot continue to increase without violating fundamental conservation laws, but how the increase becomes limited is unclear. It is hoped that the densities of quarks and gluons inside the proton will eventually be calculable from the basic principles of QCD in the same way that the densities of electrons in atoms can be calculated from the principles of electricity and quantum mechanics.

Another puzzle involves the proton's internal rotation, or spin. How the total proton spin is built up from the spins of constituent quarks and gluons must be a consequence of QCD. Naively, this spin would be formed from the principal quark constituents of the proton; surprisingly, experiments show these quarks to contribute only about 20% of the proton spin. Experimental studies of the spin structure of the proton are being done at many laboratories, including SLAC, CERN, DESY, and the new nuclear physics facilities at Brookhaven and the Continuous Electron Beam Accelerator Facility (CEBAF). Again, the challenge is to someday calculate the spin properties of the proton from QCD.

The new facilities at Brookhaven and CEBAF will also be able to explore new phases of matter that result from the strong force. It is not understood under what conditions the neutrons and protons in the nucleus dissolve into their constituent quarks and gluons. Experiments of the next decade will start to address the nature of quark gluon confinement and try to understand how the early universe evolved from a quark gluon plasma to the nuclear matter that makes up most of the visible mass in the universe today.

In parallel with the important new experimental information that is being obtained, new calculational tools are being developed and specialized parallel processing computers are being designed and constructed to perform QCD calculations. These techniques have already provided important information about the theory, and there is promise for much more.

The examples cited above illustrate that many aspects of QCD are still unexplored, both experimentally and theoretically. The QCD force underlies the bound microscopic states (protons and other atomic nuclei) of the universe's known matter, as well as the wealth of particles observed in high-energy experiments. A combination of imaginative experimental exploration, theoretical insight, and innovative calculational tools is providing a deeper understanding of QCD. There may also be exciting consequences such as new types of matter, and experimental searches for exotic states will continue at high-energy and nuclear physics facilities throughout the decade.

ARE THERE UNEXPECTED PHENOMENA?

For each of the currently operating accelerators, as well as those under construction, a full experimental program is planned. The motivation of each experiment is quite explicit, usually focusing on the issues addressed in this chapter. However, experimenters are always on the lookout for the unexpected. Although answers to questions posed incrementally can increase knowledge, the observation of an unanticipated phenomenon can revolutionize the way physicists and eventually society think about the universe. The original electron-scattering experiment revealing for the first time that the proton had an internal structure made of quarks; the discovery that a fundamental symmetry of nature, CP, was violated; and the discovery of the charm quark, which introduced a second generation of quarks—these are three spectacular examples of experiments that profoundly influenced the field with unanticipated results. In fact, many of the most important scientific discoveries have been unexpected. In experiments under way, as well as at facilities now being designed, physicists are keenly aware of this lesson. Experimenters will continue to study their data for signs of the next surprise that could significantly alter views of nature.

SUMMARY

The experiments beginning during the coming decade in this country and abroad will explore very important issues in elementary-particle physics. Researchers stand to learn a tremendous amount about the fundamental differences between matter and antimatter, enough so that the current understanding of CP violation should be either confirmed or refuted. Experiments should be able to establish generation-changing interactions among the leptons, and if so, this has important conclusions for the ultimate unification of all elementary particles. Data on the strong force will give new insights into QCD and could establish an important new phase of matter. The Higgs boson and/or some of the expected states in supersymmetry should be discovered, if they exist. If not, then physicists expect compelling evidence for a new force in nature.

Even so, one wonders what questions will remain after the coming decade of experimentation: This subject is treated after the accelerators and detectors used in particle physics are described in Chapter 6.

6

❖

Accelerators and Detectors: The Tools of Elementary-Particle Physics

INTRODUCTION

A variety of tools are required to perform the observations that form the basis of our current understanding of elementary-particle physics as developed through the twentieth century and presented in the preceding chapters. In this chapter the tools that elementary-particle physicists use to observe their world—particle accelerators and particle detectors—are described. For the past 60 years these devices have been the instruments of choice for research on elementary particles. During this period the historic advances in elementary-particle physics have been inextricably driven by advances in accelerator and detector technologies.

Elementary-particle physics (EPP) is distinguished from other accelerator-based sciences by its reliance on accelerators operating at the highest energies attainable with present technology—the "energy frontier." To sustain continued progress in elementary-particle physics it is necessary to create conditions under which elementary particles—protons, electrons, muons, neutrinos (and their antiparticles)—interact at extremely high energies and in quantities sufficient to allow observation of extremely rare processes. Such conditions do not exist naturally on Earth, and cosmic rays of sufficiently high energy are too rare. This

---

Note: This chapter contains more technically detailed information than other chapters in the report. The interested reader is encouraged to read the chapter in its entirety. However, the sections "Performance of Existing Accelerators," "Accelerator Facilities Under Construction," and "Particle Detector Topologies" can be omitted without missing the essence of this report.

need has driven the construction of ever larger accelerator facilities and increasingly large and complex particle detectors.

The two most important properties characterizing the utility of a facility for EPP research are energy and luminosity. Today, physicists are able to produce collisions between elementary particles with energies close to 1 TeV (10^{12} electron volts)—the energy equivalent of 10^{12} V energy source. This represents a millionfold increase since the invention of the cyclotron 65 years ago (the difference in scale is shown in Figure 6.1). Until the 1960s, accelerator-based elementary-particle experiments relied exclusively on directing particle beams onto bulk matter—the so-called stationary target configuration. However, over the past 30 years, energy performance has been greatly enhanced by the development of the "particle collider"—an accelerator configuration in which particles and/or antiparticles collide head-on. As described in Chapter 2, the collider configuration provides the most efficient mechanism for translating beam energy into collision energy and thus provides the most direct access to the energy frontier.

The luminosity of a facility is a measure of the rate at which particles collide. Luminosity is directly related to the intensity of the particle beam (or beams) employed (and, in a collider, to the size of a spot onto which the beams are focused). Elementary-particle physicists measure luminosity in units of inverse square centimeters times inverse seconds ($cm^{-2} s^{-1}$). This allows one to calculate an event rate by multiplying luminosity by the effective cross-sectional area of the particles that are colliding. Typical luminosities are in the range 10^{30} to 10^{35} $cm^{-2} s^{-1}$. Such large luminosities are required because the effective areas of the colliding particles are so small—for example, high-energy electrons have an effective area for producing Z^0 particles of only about 10^{-32} cm^2, leading to an interaction rate of one every 10 seconds in a facility operating with a luminosity of 10^{31} $cm^{-2} s^{-1}$.

The highest luminosities are attained at stationary target facilities where one can use a dense solid as a target. At the Brookhaven Alternating Gradient Synchrotron (AGS), for example, several trillion protons can be made to interact with stationary targets every second. Stationary target facilities are often used to produce intense beams of particles that can not be accelerated or stored in collider facilities because of their short lifetimes (muons, charged K and p mesons) or their lack of electric charge (neutrons and neutrinos).

Luminosity at collider facilities is much lower due to the low density of the beams relative to ordinary matter. However, for observations undertaken at these machines, the increased operating energy more than compensates for the lower luminosity. At the Fermilab Tevatron, for example, current operations produce about 500,000 annihilations every second between protons and antiprotons in the two countercirculating beams. Experimenters were recently able to identify approximately 100 examples of production of the top quark based on a year of operations at this facility. The observation and measurement of such extremely rare processes requires highly sophisticated particle detectors supported by state-of-the-art computer facilities.

Particle beams contain anywhere between 10^{10} and 10^{14} particles. Although the total mass contained in these beams is minuscule, less than one ten-billionth of a gram, the energy is contained within an incredibly small volume. Beam sizes range from the size of a human hair, in the Fermilab Tevatron, to a hundred times smaller in the Stanford Linear Collider (SLC) at SLAC.

The detectors of elementary-particle physics are used to observe the debris produced in a single collision of two individual particles—an "event." Such events often produce up to a hundred individual particles emanating from a small region surrounding the "interaction point." The job of the detector is to characterize an event in terms of the energy (or momentum) and type of particles produced. This characterization is based on reconstruction of the tracks, or footprints, left by the particles in the detector. After reconstruction of an event, the elementary-particle physicist must interpret the event in terms of the underlying physical processes involved. An example of an event in which a pair of top quarks are produced has been shown in Figure 4.6.

FIGURE 6.1 (a) The 11-inch cyclotron in a Berkeley laboratory in the early 1930s, in contrast with (b) a sector of the Tevatron in the mid-1980s. (Courtesy of the Lawrence Berkeley National Laboratory and the Fermi National Accelerator Laboratory.)

Over the preceding decades, the need for ever higher energy has led to the construction of ever larger (and more expensive) facilities. As cost and complexity have increased, the number of such facilities has decreased to the point that today there are nine major elementary-particle physics laboratories in the world. These laboratories and their operating facilities are listed in Table 6.1.

The highest-energy collider operational in the world today is the Fermilab Tevatron, situated in Batavia, Illinois. This facility accelerates protons and anti-protons to 900 GeV (1 GeV = 10^9 eV) per beam and is based on superconducting magnet technology. The Tevatron supports operations in both collider and stationary target mode and will remain the highest-energy facility in the world until the initiation of operations at the Large Hadron Collider (LHC) facility in Geneva, Switzerland, currently expected in the middle of the next decade. The SLC in Palo Alto, California, is an electron-positron ($e^+ e^-$) collider capable of producing collisions at 45 GeV per beam. It is one of two facilities in the world capable of direct production of the Z^0 boson and the only one with a polarized electron capability. The SLC is the first facility ever built as a "linear collider"— the technology that appears to be required for further extension of the energy of electron-based facilities. Other facilities in the United States include the Cornell Electron Storage Ring (CESR), an electron-positron collider operating at approximately 5 GeV per beam for the study of B mesons (the highest-luminosity

TABLE 6.1a Elementary-Particle Physics Facilities Operational in the World Today—Collider Facilities

Laboratory, Facility	Beam Energy (GeV)	Center-of-Mass Energy (GeV)	Particle Types	Luminosity ($cm^{-2} s^{-1}$)	Start of Operations	Location
Fermi National Accelerator Laboratory, Tevatron	900	1,800	Proton-antiproton	2×10^{31}	1986	Batavia, Illinois
Stanford Linear Accelerator Center, SLC	45	90	Electron-positron	1×10^{30}	1989	Palo Alto, California
Cornell University, CESR	5	10	Electron-positron	4×10^{32}	1980	Ithaca, New York
CERN, LEP	91	182	Electron-positron	3×10^{31}	1989	Geneva, Switzerland
KEK, Tristan[a]	32	64	Electron-positron	1×10^{31}	1986	Tsukuba, Japan
Institute of High Energy Physics, BEPC	2	4	Electron-positron	6×10^{30}	1989	Beijing, China
Budker Institute of Nuclear Physics, VEPP-2M	0.7	1.4	Electron-positron	3×10^{31}	1989	Novosibirsk, Russia
DESY, HERA	30 (electrons) 820 (protons)	310	Electron-proton	3×10^{30}	1991	Hamburg, Germany

[a] Operations at Tristan ceased in December 1995.

TABLE 6.1b Elementary-Particle Physics Facilities Operational in the World Today—Stationary Target Facilities

Laboratory, Facility	Beam Energy (GeV)	Center-of-Mass Energy (GeV)	Particle Type	Beam Intensity (particles/sec)	Start of Operations	Location
Fermi National Accelerator Laboratory, Tevatron	800	39	Proton	4×10^{11}	1983	Batavia, Illinois
Brookhaven National Laboratory, AGS	30	8	Proton or heavy ion	3×10^{13}	1960	Upton, New York
Stanford Linear Accelerator Center, LINAC	50	10	Electron	1×10^{13}	1967	Palo Alto, California
CERN, SPS	440	29	Proton	4×10^{12}	1976	Geneva, Switzerland
KEK, KEK-PS	12	5	Proton	2×10^{12}	1976	Tsukuba, Japan

collider ever built), and the AGS at Brookhaven National Laboratory, a 30-GeV proton facility utilized for a variety of high-precision measurements, with the highest beam intensity available for elementary-particle physics in the world. Major facilities overseas include the Large Electron Positron collider (LEP) at CERN (the European Laboratory for Particle Physics), the highest-energy $e^+ e^-$ collider ever constructed; the electron-proton collider, HERA, at the DESY laboratory in Hamburg; and modest-energy electron colliders in Beijing and Novosibirsk. (The facility at the KEK laboratory in Japan is currently being upgraded, as discussed later in this chapter.)

PARTICLE ACCELERATORS

The term "particle accelerator" refers generically to a machine in which elementary particles can be accelerated and/or stored. A "collider" is a special kind of accelerator in which particle beams are directed at each other, producing head-on collisions. In all accelerators constructed to date, the accelerated particles are both stable and electrically charged, limiting the possibilities to protons, electrons, and their antiparticles (although ions are often used in applications outside EPP). Fortunately, it is possible to create "particle beams" of a variety of short-lived or neutral particles. Such particle beams are usually produced through interaction of a high-energy proton beam with a stationary target, resulting in a "secondary particle beam," and are a standard feature of proton-based stationary target facilities.

Protons and electrons used in particle accelerators are relatively easy to produce. Protons are extracted from ionized hydrogen gas, whereas electrons are emitted from a cathode either under the influence of heating (as in a television or computer monitor) or under the influence of incident laser light. Antiprotons and antielectrons (also known as positrons) do not occur naturally and so have to be created. Both are produced by directing particle beams onto pieces of ordinary bulk matter. In practice it is much easier to produce positrons than antiprotons because of the much lower mass of the positron. By interacting electrons with lead it is possible to produce approximately one positron for every electron, whereas it takes approximately 60,000 protons interacting in a nickel target to produce a single usable antiproton.

Performance of Existing Accelerators

Modern accelerators used in support of elementary-particle physics research come in two basic types: linear accelerators (linacs for short), and synchrotrons. Both types trace their lineage back to the late 1920s and early 1930s. It is remarkable that although both the size and the technological basis of these machines have changed dramatically over the past 60 years, the basic operating principles remain the same—particle beams are accelerated by electric fields and confined by magnetic fields.

Linear accelerators run in straight lines. The energy of the beam delivered from a linac is simply the product of the average applied electric field and the length. The largest linac built to date is the SLAC linac, capable of accelerating electrons to 50 GeV. This device is roughly 3 km (2 miles) long, having an average accelerating field of about 17×10^6 V/m (see Figure 6.2 in color well following p. 112). Increasing the energy performance in linacs depends on raising the electric field that can be applied to the beam. Accelerating systems currently under development are aimed at doubling to tripling the accelerating fields in these facilities.

Circular accelerators are descendants of the early cyclotrons and utilize magnetic fields to bring particles around a circle so that the accelerating electric field can be applied repeatedly to the beam to increase its energy. In a synchrotron, the recirculation circle is fixed, so the confining magnetic field must be increased in direct proportion to the energy of the beam. In this case the energy of the synchrotron is the product of the number of revolutions a particle makes around the accelerator and the average accelerating voltage seen on each revolution. The Fermilab Tevatron and the Brookhaven AGS are synchrotrons. The Tevatron achieves its operating energy of 900 GeV through the repeated application of an accelerating voltage of 10^6 V over approximately a million revolutions of the machine.

In principle, the energy could be increased indefinitely in a synchrotron by continued application of the accelerating voltage over many revolutions, but the product of the confining magnetic field and accelerator radius must be high enough to keep the particle beam circulating at the highest energy. The maximum energy attainable in proton synchrotrons has been increased most recently through the application of superconducting magnet technology. This technology has allowed more than a doubling of the magnetic fields that can be produced and hence more than a doubling of the attainable energy for a fixed-circumference ring. The Fermilab Tevatron was the first high-energy accelerator ever built utilizing superconducting magnets. Tevatron magnets operate at 44,000 gauss (G). (By way of reference, the magnetic field of the Earth is about 0.5 G and the largest field achievable in a room-temperature electromagnet is about 20,000 G.) Magnets developed for the canceled Superconducting Super Collider (SSC) project operated at 66,000 G, whereas those under development for the LHC project in Europe are specified at 90,000 G.

In contrast, the energy of electron synchrotrons is limited not by the ability to produce large magnetic fields but by the need to replenish power radiated in the form of synchrotron light. This phenomenon, known as "synchrotron radiation," occurs because high-energy electrons lose energy by radiating light when forced to travel in circles. The effect is so strong that every doubling of the electrons' energy is accompanied by a sixteenfold increase in radiated power.

However, this effect is reduced in direct proportion to the size of the circle traced out by the electron. For an accelerator such as LEP the radiated power is between 10 and 20 MW—about the power consumption of a large town. To keep the power manageable, the radius chosen is 4.2 km, and the magnet field required to guide a 90-GeV electron beam is a mere 720 G. It is illustrative of the difference between electron and proton synchrotron capabilities that this same tunnel will support a 7,000-GeV proton beam once the LHC accelerator has been installed. LEP is the largest particle accelerator in the world today. The very strong energy dependence of synchrotron-radiated power is reason to believe that it will be difficult to extend further the energy frontier for electrons by utilizing circular accelerators and storage rings. This belief has provided the motivation for construction of a colliding linac, the SLC, as the first example of the technology that will be required to extend the electron energy frontier.

It is evident that much higher energies are achievable in proton accelerators than in electron accelerators. This situation is expected to persist into the foreseeable future. As explained in Chapter 2, electron accelerators remain competitive because of the more fundamental nature of the electron than the proton. Accompanying luminosities are similar in the two types of accelerators.

Accelerator Facilities Under Construction

Continued pursuit of elementary-particle physics requires access to facilities of continually improving performance. A number of upgrades to existing EPP facilities are currently under way in the United States and abroad aimed at supporting the research needs of the elementary-particle physics community over the next 10 years. These projects are listed in Table 6.2. Upgrades to existing facilities are highly cost-effective, building on the significant investments in the "base" laboratory. However, it should be noted that with the exception of the LHC, each of the projects listed targets increased luminosity; only the LHC provides a significant extension of the energy frontier, as well.

The Main Injector project at Fermilab involves construction of a new rapid-cycling 150-GeV proton accelerator that will support an increase in the intensity of the proton and antiproton beams in the Tevatron. The goal is to boost Tevatron collider luminosity by a factor of five beyond current operations. Additionally, a new antiproton storage ring, known as the Recycler, has been recently incorporated into the project. This ring holds the promise of a further increase of a factor between 2 and 10 in luminosity. A secondary benefit of the Main Injector project will be the creation of a new capability for high-intensity stationary target operations at 120 GeV. The Main Injector accelerator is based on conventional, room-temperature, magnet technology, whereas the Recycler is designed to utilize permanent magnets—the first large-scale use of permanent magnet technology in a storage ring.

TABLE 6.2a Major Upgrades and New Facilities Under or Approved for Construction—Facilities Upgrades

Laboratory, Project	Operational Goal	Start of Operations	Location
Fermi National Accelerator Laboratory, Main Injector	Proton-antiproton collisions at 2,000 GeV and 2×10^{32} cm^{-2} s^{-1} 120-GeV protons for stationary target operations	1999	Batavia, Illinois
Stanford Linear Accelerator Center, PEP-II	Asymmetric electron-positron collisions at 10 GeV and 3×10^{33} cm^{-2} s^{-1}	1999	Palo Alto, California
Cornell University, CESR	Symmetric electron-positron collisions at 10 GeV and 2×10^{33} cm^{-2} s^{-1}	1998	Ithaca, New York
CERN, LEP-II	Electron-positron collisions at 192 GeV and 1×10^{32} cm^{-2} s^{-1}	1998	Geneva, Switzerland
KEK, *B* Factory	Asymmetric electron-positron collisions at 10 GeV and 3×10^{33} cm^{-2} s^{-1}	1999	Tsukuba, Japan

TABLE 6.2b Major Upgrades and New Facilities Under or Approved for Construction—New Facilities

Laboratory, Project	Operational Goal	Start of Operations	Location
Brookhaven National Laboratory, RHIC	Polarized proton-proton collisions at 500 GeV and 2×10^{32} cm^{-2} s^{-1}	2000	Upton, New York
Frascati, DAPHNE	Symmetric electron-positron collisions at 1 GeV and 5×10^{32} cm^{-2} s^{-1}	1998	Frascati, Italy
CERN, LHC	Proton-proton collisions at 14,000 GeV and 1×10^{34} cm^{-2} s^{-1}	2005	Geneva, Switzerland

The PEP-II project at SLAC is an upgrade of the original Positron-Electron Project (PEP) electron-positron colliding beam facility to support investigation of CP violation in the *B*-meson system. The facility is based on two rings operating at different energies, roughly 3 and 8 GeV. Technology challenges are related to the very high circulating currents requiring state-of-the-art feedback systems. This facility will also represent the first operation of an $e^+ e^-$ colliding beam facility with unequal beam energies. A similar facility is under construction at KEK in Japan.

Cornell is also upgrading the very successful CESR electron-positron collider to achieve higher luminosity, again for detailed study of the *B*-meson system. In contrast to the PEP-II facility, the CESR upgrade will continue to utilize equal energy beams circulating in a common storage ring.

A new colliding beam facility, the Relativistic Heavy Ion Collider (RHIC), is under construction at Brookhaven National Laboratory, with operations scheduled to commence in 1999. This facility is being funded as a nuclear physics project and will have as a primary operational mode the collision of heavy ions, up to fully ionized Au^{+79}. RHIC will also have a capability for producing collisions between polarized protons at 500 GeV in the center of mass. The RHIC facility is based on superconducting magnet technology.

Overseas, a significant upgrade to the energy of the LEP collider is proceeding with a goal of achieving 96 to 100 GeV per beam by 1998 to 1999. However, the major initiative worldwide is the LHC. This facility, with seven times the energy of Fermilab, is approved for construction at CERN. With initiation of operations at the LHC projected for the year 2005, the energy frontier will move from Fermilab in the United States to CERN in Europe. The LHC is designed to collide two countercirculating proton beams at a center-of-mass energy of 14 TeV. The LHC will be constructed within the existing 26 km LEP tunnel and is based on the superconducting magnet technology first developed at the Fermilab Tevatron and improved on at Brookhaven, HERA, and the SSC.

Options for Future Facilities

A variety of projects that could extend the energy frontier up to or beyond the LHC are in various stages of development in the United States and abroad. Possibilities include electron-positron colliders operating in the range of 0.5 to 1.5 TeV, muon colliders operating in the range of 0.5 to 4.0 TeV, and very large hadron colliders with energies in the range of 50 to 100 TeV. Any such facility would be required to support a factor of about 100 increase in luminosity relative to current achievements. These possibilities are summarized in Table 6.3. All facilities listed are likely to have associated construction costs measured in excess of $1 billion. Cost efficiencies in manufacturing, installation, and operations are a major concern of the development efforts for each.

TABLE 6.3   Potential Major Facilities to Be Constructed After the Year 2005

Facility	Particle Types	Energy (center of mass)	Enabling Technologies
Next-generation linear collider	Electron-positron	0.5-1.5 TeV	Microwave power sources and accelerating structures. Cost-efficient manufacturing
Muon collider	Muon-antimuon	0.5-4.0 TeV	Superconducting magnets and accelerating structures. Beam cooling. High-intensity particle beams and targeting
Next-generation large hadron collider	Proton-proton	50-100 TeV	Superconducting magnets. Cost-efficient manufacturing and tunneling

Research and development programs aimed at new electron colliders are by far the most advanced of the areas listed above. Because of the relationship between radiated power, energy, and accelerator size, all electron collider designs currently under consideration are based on the linear collider configuration pioneered by SLAC. Most effort to date has been devoted to investigating the 0.5 to 1.0 TeV range. Any facility operating in this range is expected to have a linear extent of 30 to 50 km. Major R&D efforts are currently centered at SLAC in this country, KEK in Japan, DESY in Germany, and CERN in Switzerland. Communication and cooperation between the major development centers has been a significant feature of the R&D program for many years and is expected to lay the groundwork for international cooperation on a next-generation linear collider when and if one is built.

The SLAC and KEK approaches to linear colliders are quite similar—both have developed designs based on room-temperature copper-accelerating structures driven by high-efficiency radio-frequency power sources known as klystrons. The DESY approach differs in that it relies on a superconducting accelerating structure, and CERN is exploring accelerating structures driven by a dedicated electron "drive" beam. The warm and superconducting approaches are complementary. Room-temperature designs require very small (a few nanometers in height) beam sizes and high bunch current. Component fabrication and alignment tolerances, precision control of beam trajectories, and removal of optical aberrations to high order are critical in these designs. Many of these tolerances are relaxed in the superconducting design. The trade-off is that super-conducting accelerating structures are inherently more expensive and provide a lower accelerating gradient, thus requiring a facility nearly twice as long as the room-temperature structures. Proponents of each of these approaches are now pursuing integrated conceptual designs and cost estimates. The expectation is that such designs could be available in the next 3 to 5 years.

A significant effort directed toward developing design concepts for a muon collider has grown over the past several years in the United States. The muon collider is a facility in which muons are produced, accelerated, and collided at very high energies. No such facility has ever been built. The physics research supported by a muon collider would be similar to that supported by an electron collider. The virtue of the muon is that with a mass 200 times that of the electron, synchrotron radiation effects are greatly suppressed, affording the pos-sibility of operations at higher energies. The downside is that muons are not stable—once produced they decay in a few thousandths of a second, leaving little time to accelerate and bring them into collision. A feasibility study has been developed for a facility operating in the 0.5 to 4.0 TeV range. Significant issues are identified as requiring solution before a viable design for such a facil-ity can be contemplated. Many of the required developments are symbiotic with R&D efforts on other projects. For example, proton targeting for muon produc-tion shares many issues in common with antiproton or neutron production, whereas the muon-accelerating structures required are very similar to those be-ing developed for the linear collider effort at DESY. With sufficient support, development of a complete conceptual design for such a facility, if one can be built, could probably be forthcoming in the period 2005 to 2010.

Scientists at a number of U.S. laboratories and universities have started to identify requirements for a new proton-proton collider with an energy reach a factor of approximately 10 beyond LHC. Although current hadron collider tech-nology could probably form the basis for such a facility, the SSC experience has convinced those involved that significant reduction of construction and operat-ing costs is absolutely critical. Two approaches are currently under study, one based on relatively low- and one on very high-field superconducting magnets. Both approaches require advances in superconducting material and magnet tech-nologies, as well as advances in tunneling technologies beyond the current state of the art in order to achieve desired cost efficiencies. Such superconducting magnet R&D represent a natural outgrowth of existing programs at Fermilab, Brookhaven, and Lawrence Berkeley Laboratory, all of whom are currently en-gaged in R&D in support of the LHC construction. Close cooperation with the domestic and foreign superconducting materials industry is expected to be mutu-ally beneficial throughout this development period. It appears likely that, spurred by recent rapid advances in high-temperature superconducting technologies in industry, the next hadron collider beyond LHC will utilize these new developments. With sufficient developmental support, a viable conceptual design for a proton-proton collider at least seven times more powerful than LHC could probably be developed sometime from 2005 to 2010.

All the efforts described here require significant R&D to move forward. Lead times associated with the design and construction of new accelerators have become long enough that R&D into promising areas is required now if the United States is to be in a position to participate in, or host, a new forefront accelerator facility in the post-LHC era. R&D in support of a second-generation linear collider design is relatively advanced at this time but has to be continued through cost minimization studies and the development of a complete conceptual design. R&D in support of very large hadron colliders and muon colliders is in the early stages of development and has to be nurtured. Significant support will be required in the areas of superconducting materials, superconducting magnet design, superconducting accelerating structures, development of very high intensity proton beams, and ionization cooling if hadron or muon collider concepts are to be developed to a degree that will offer the United States viable choices for continued leadership in elementary-particle physics into the extended future.

DETECTORS IN ELEMENTARY-PARTICLE PHYSICS

The role of the accelerators described earlier is to bring high-energy particles into collision at a well-defined point in space—the interaction point. The interaction point is viewed by a detector whose job is to take a snapshot of the debris produced in the collision. Events are observed at rates up to several million per second in some experiments, and each event can involve the production of hundreds of particles. Of these collisions, only a very few are interesting and represent a phenomenon that has not been frequently observed before.

The frequency of collision and the volume of information required to characterize individual events constrain the methods of detection that can be employed. Many years ago it was possible to perform experiments in which the human played the primary role in the acquisition and storage of data. The physicist could observe, by eye, some flashes of light, indirectly from nuclear disintegrations, and then make some notes in a logbook. In modern detectors, such techniques have, by necessity, given way to a vast array of high technology often involving millions of the most advanced electronic circuits and the highest-performance computers available. The strategies involved in producing the event snapshot are discussed below. In general, the approaches utilized are common to lepton collider, hadron collider, fixed-target, and cosmic-ray experiments.

The snapshot ideally contains sufficient information to characterize any given event. The complete picture may be built up from detection of all the individual elementary particles in the event. These "elementary" particles may be discrete particles or they may be jets of hadronic particles. Sometimes the presence of a particle can be inferred only by invoking rather general laws of physics, such as momentum conservation, to determine what is "missing." An example is the original evidence for the existence of the neutrino.

Most particles produced in an interaction are long-lived, whereas a few may decay before they exit the detector, allowing observation of the decay products as an aid to identification of the parent particle. As an example, hadrons containing *b* quarks travel just a few millimeters before decaying. The decay vertex can be distinguished from the primary vertex by using very high resolution silicon-based detectors.

A given experiment is a collection of different detectors that, when assembled together about the interaction region, gives a complete picture of an event. Such assemblies are never perfect—any given design has its own strengths and weaknesses. Choices are driven by the physics interests. The individual building blocks can be employed in different environments in different ways, but the components of lepton collider experiments and hadron collider experiments are more remarkable for their commonality than their differences.

Particle Detection

The first step in constructing the event snapshot is to identify the presence of individual particles and to measure their tracks. The basis for detecting the presence of an individual particle is the exploitation of some well-understood physical process, typically the electromagnetic interaction of a charged particle with matter. For example, when Henri Becquerel discovered in 1896 that uranium was radioactive, he saw light traces in photographic paper. The light was produced by the ionization and subsequent de-excitation of some atoms by the passage of charged particles produced in the radioactive decay of uranium. This process is called scintillation and remains a very common detection method. If the same ionization process is induced by the passage of a charged particle through a gas in which there is a strong electric field, the liberated electron is accelerated in the direction of the field, and by collision with another atom, can produce a second ionization. The process can multiply into an avalanche of electrons. The result is an electrical impulse that can be detected.

An early example of the detection of individual particles using the ionization process is the familiar Geiger-Müller tube. A more modern example is the silicon strip detector. The understanding of the beautiful properties of semiconductors is responsible for the revolutionary advances that have made today's electronics industry possible. Over the last 15 years, new multichannel semiconductor detectors have also transformed major high-energy experiments by allowing a tenfold increase in the precision with which particle tracks can be identified.

The basic silicon strip detection device is a silicon diode. The material properties of silicon are such that the passage of a charged particle promotes electrons to the conduction band of the material in a manner analogous to the ionization of a gas. The electrons and their corresponding hole pairs are then free to drift under the influence of an electric field. Thus, in the presence of a modest voltage applied between electrodes on either side of the silicon wafer, an electric pulse appears upon passage of a charged particle. What has made this type of detector so attractive for particle physicists is the ability to implant the appropriate electrodes with a separation of about 50 μm, allowing the position of a charged particle to be determined with a precision of about 10 μm. Figure 6.3 (in color well following p. 112) shows an example of particle tracks emanating from a particle interaction as captured by a silicon detector.

Other particle detection techniques are also frequently employed, two examples being Cerenkov detection and calorimetry. Cerenkov light is a phenomenon rooted in the fact that a particle can travel with a velocity greater than the speed of light in a medium—for example, a gas—that has a refractive index greater than 1. When this happens, the optical equivalent of a sonic boom is produced and light is emitted. Thus, the presence or absence of a light burst when a charged particle traverses a gas indicates whether the velocity of the particle is higher or lower than a certain value. Calorimetric detection methods, on the other hand, rely on attempting to have a particle interact with some material in such a manner that all of its energy is deposited in a measurable way. The generic name for devices utilizing this technique is "calorimeters." Calorimetry has the virtue of working equally well for particles with and without an electric charge.

Modern particle physics detectors are assembled from a wide array of individual detection elements. The elements each have different purposes and different capabilities. Some measure particle trajectories, others measure velocity, and still others measure total energy. Some can see neutral particles and some cannot. Often, information from a variety of detector elements is used to deduce the identity of a particle. As described below, physicists assemble detectors in a manner that allows for coordinated use of the information collected from all the detection elements to form the most accurate picture of the event possible.

Particle Detector Topologies

The goal of physicists assembling a variety of individual detection elements into an integrated detector is to obtain an accurate snapshot of an event. This usually means measuring the momentum and ascertaining the identity of as many of the particles produced in the event as possible. The individual techniques for particle detection and measurement utilized in modern particle detectors are broadly applicable to all types of experiments. However, the topologies of different detectors often differ considerably. In stationary target experiments, because of the high momentum of the beam particles, the particles produced generally tend to move in a small cone in the direction of the incident beam. Stationary target detectors therefore tend to be restricted to cones about the beam direction. A typical stationary target detector might measure 3 m in diameter and 100 m in length. In collider experiments, by contrast, the two colliding beams are usually of equal energy. Forward and backward tend to be equivalent, so there is a tendency to use a detector that extends approximately the same length in all directions, typically 5 to 10 m. Intermediate between these two extremes are detectors at colliders in which the beams are of unequal energy. There is a desire for coverage of all directions, with a strong emphasis on the direction of the higher-energy beam. Beyond this topological consideration, all detectors are very similar.

One of the striking aspects of a particle physics detector is the range of scales involved. Detectors for new experiments at the LHC at CERN will measure 10 m on a side—1,000 m³. Their weight will be tens of thousands of tons. In contrast, embedded in the heart of the detector will be devices 3×10^{-4} m thick, capable of measuring position with a precision of 1×10^{-5} m. In distances, the range of scales is 1 million!

Figure 6.4 shows a typical topology for a detector utilized in a colliding beam experiment. The integration of a variety of individual detection elements into a large detector represents a particular strategy for creating the event snapshot. By starting from the point of interaction and moving outward, it makes sense to determine a particle trajectory before the particle is deflected or absorbed through interaction with dense matter. Nearly all detectors measure particle momenta with tracking chambers immersed in a magnetic field. In the generic design shown, a solenoid magnet is placed so as to enclose the tracking volume inside the particle calorimeters. The calorimeters are near the outside because the particles, with a few exceptions as noted below, are completely absorbed in these devices.

Layer 1: Charged-Particle Tracking Volume

Large arrays of wire or silicon ionization detectors can be assembled to measure the trajectories, or tracks, of individual particles. Trajectories are used to determine the point of origin and momentum of the particles. Closest to the primary interaction point, one usually deploys a series of layers of the highest-precision detectors possible. Typically these are silicon strip detectors. These detectors are used to identify unstable particles that travel a short distance from the interaction point before transmuting into different particle types (e.g., *B* mesons) as shown in Figure 6.3. The precision with which a secondary vertex can be distinguished from the primary vertex is determined by the closeness of the

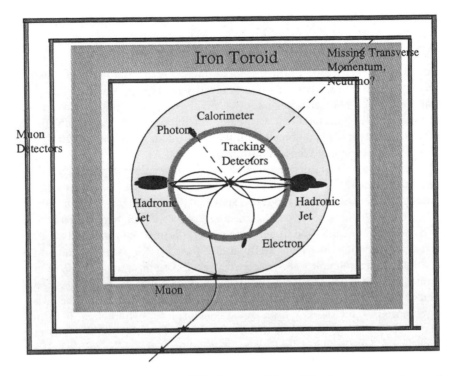

FIGURE 6.4 Figure of a generic collider detector with particle signatures superimposed.

innermost detector to the interaction point and the measurement precision of the detector.

Outside the silicon detectors, the remainder of the tracking volume is filled with detectors of moderately high precision immersed in a magnetic field. The goal here is to measure the momentum of individual charged particles by measuring the trajectories as they traverse the magnetic field. Momentum is calculated by using the property that a charged particle travels on a circular trajectory in the presence of a magnetic field. The radius of curvature of the orbit is directly proportional to the momentum of the particle and inversely proportional to the strength of the magnetic field. Unfortunately, the resolution of this technique decreases as the particle momentum increases.

As might be expected, the more precisely the trajectory is measured, the more accurately is the momentum determined. This leads to choices in construction inevitably resulting in trade-offs between resolution and cost. Because of the desire to keep particles following their naturally curved trajectories, every attempt is made to eliminate extraneous material that might deflect the particles, and thereby diminish measurement accuracy, as they traverse this region.

Layer 2: Charged-Particle Identification

The techniques described above allow detection of the presence of individual charged particles and measurement of their momenta and location in space. Complete characterization of the event also requires the ability to determine the identity of the charged particles and to identify the presence of electrically neutral particles. A number of techniques are employed to fulfill this need. In both of the examples presented below, the information obtained from a specific type of particle detector is combined with the momentum information generated by the tracking detectors to lead to information on particle identity.

The speed of a charged particle of a given momentum depends on its mass. The most straightforward method for determining the speed of a particle is to measure the time it takes to traverse a specified distance in the detector. Because of the high velocities involved, measurement of the elapsed time with a precision of a few 10^{-10} s is required to generate useful information. Such a technique is referred to as "time of flight" and is useful for particles with momenta up to a few billion electron volts; above this energy, all particles tend to have velocities indistinguishable from the speed of light, independent of their identity. Generation of Cerenkov light can also be used as an aid to identification. The presence or absence of the Cerenkov light signal contains information about the velocity of a particle and hence can help distinguish between particles of different masses such as the electron, muon, or proton. Again, this technique is viable only up to momenta of several billion electron volts.

Layer 3: Calorimetric Energy Measurement

Electrons are very light particles, and their electromagnetic interactions with high atomic number materials, such as lead, result in a shower of electrons, positrons, and photons as first the parent and then its daughter particles emit photons. Photons, in turn, produce pairs of electrons and positrons and so on. This shower development and the manner in which the original energy of the electron is deposited in material provide a characteristic signature of the electron. The total energy deposited in the calorimeter is that of the original particle.

Photons, the particles of light, develop showers in almost exactly the same way as electrons. Thus, their calorimetric signature is the same as that of electrons. However, being electrically neutral, photons can be distinguished from electrons—photons leave no track in the charged particle tracking volume whereas electrons will.

Hadrons—typically pions, kaons, neutrons, and protons—can interact with matter through strong nuclear interaction. Electrons, positrons, and muons do not possess this capability. Therefore hadrons produce showers with different characteristics. Given the difference in interactions and in the properties of materials, hadronic showers are more extensive than electromagnetic showers both longitudinally and laterally. Thus, the pattern of energy deposition distinguishes them from both electrons and muons. Again, as with an electron, the total energy deposited is that of the original particle.

Calorimeters have several virtues. As mentioned earlier, they are sensitive to both electrically charged and electrically neutral particles. Calorimeters are also a very effective means of identifying and measuring the energy contained in jets associated with quark production as described in Chapter 4. A rather nice feature of the measurements is that, in contrast to tracking measurements, the energy resolution improves as the particle energy increases. Calorimeters are heavy, large, and expensive, often dominating the overall cost of a modern detector.

Layer 4: Muon Identification and Measurement

The massive calorimeter is the end of the line for most particles. All the photons, electrons, and hadrons interact therein, deposit all their energies, and proceed no further. The two particles to which this does not apply are the muon and the neutrino. Muons are often dubbed "heavy electrons." Because of their higher mass, a muon of a given momentum will develop a much weaker electromagnetic shower than the corresponding electron. This occurs to such an extent that muons penetrate matter with relative ease, leaving only small fractions of their energy through ionization and similar processes.

A muon's momentum may have been measured in the inner tracking volume; however, at that point there is no distinction between it and other particles. Muons may be identified by observing particles that penetrate the calorimetry and match one of the tracks in the inner volume. If the determination of inner tracking momentum is adequate, muon detectors are often characterized by multiple layers of measurement separated by further layers of dense material. Sometimes, this material is iron, which may be magnetized in an attempt to enhance identification. In other cases, large magnetic field volumes may be employed, with little interleaved material, in an attempt to improve the momentum measurement of muons in a region free of confusion by other trajectories.

Hermeticity: Neutrino Measurement

Neutrinos are neutral and, as far as determined to date, massless; they interact with matter only through the weak interaction. Experiments have been performed in which beams of billions of neutrinos impinge on detectors weighing many thousands of tons with only one or two interacting in any manner. So if a neutrino is produced in the primary interaction or in subsequent decays, it invariably escapes direct detection. However, the law of conservation of momentum applies to the interactions being observed. Therefore, if all other particles produced in an interaction are measured, the absence or presence of a neutrino (or perhaps other hypothetical penetrating particles) can be inferred by comparing the total momentum of the observed interaction products to the total momentum of the beam and target particles prior to the collision. Of course, if more than one neutrino is produced in an event, the situation is more complicated. The extent to which a detector is capable of determining the existence of an energetic neutrino is therefore controlled by its ability to measure all other particles. This feature is known as "hermeticity": If a detector is hermetic, nothing escapes detection. Since closing the box to make a hermetic detector controls the extent of the detector, it also controls the cost.

Triggering

One of the most important jobs performed by the detector is to decide when an interaction is of sufficient interest to record permanently all the information contained in individual sensors and detectors for later examination (i.e., when to hit the shutter release to take a snapshot). This function, referred to as the "trigger," is carried out by a large array of sophisticated electronics circuits and computers that examine the information flowing in and decide whether the event is worthy of further consideration. Most modern elementary-particle experiments rely on hierarchical triggers, in which decisions are made at two or three different levels. Each level provides a go or no-go decision to proceed to additional, more sophisticated processing of information at the next level. Only events satisfying the highest-level trigger are permanently recorded. In this manner, fundamental interaction rates of 1 million per second can be reduced to a more manageable level of 100 events or less per second recorded permanently. With a properly designed trigger, the interesting events, which represent new phenomena, survive through the highest trigger level whereas uninteresting events do not.

Challenges for the Next 10 to 20 Years

The needs for detector research and development for the next 10 years or more are dominated by issues involved in experimentation at the LHC and, to a lesser extent, in a possible future lepton collider. In addition to building on the experience gained at existing collider facilities, many of the issues being confronted by LHC detector designers were identified, and in many cases resolved, as a result of R&D programs initiated during the SSC era.

The characteristic physics in which EPP researchers are interested has, as its origin, interactions between the fundamental fermions of the present Standard Model. The interaction probabilities decrease very rapidly as the energy increases, pushing elementary-particle physicists to demand ever higher luminosities at higher-energy accelerators. Unfortunately, the backgrounds associated with uninteresting processes do not decrease nearly as quickly. As a result, ever larger fluxes of particles associated with uninteresting interactions will flood detectors of the future, leading to potential confusion in the search for more interesting, but rare, interactions. The proverbial needle gets smaller, but the haystack gets bigger.

Occupancy

Increased luminosity means that in time and space, particles resulting from interactions are grouped more closely together. If the probability that an individual detector element will have a hit when interrogated reaches about 10%, the utility of the detector becomes problematic. There are two ways to improve the situation:

1. *Granularity.* The occupancy in a detector element can be reduced by reducing the size of the element. This is happening with silicon detectors. At modest luminosities it is possible to work with strip counters. Two stereo views can then provide the relevant two-dimensional information. As luminosities and occupancies increase, the probability of mistaken association increases dramatically. The solution is to decrease the long dimension of the strip. The result is a rectangular pixel with almost equal sides. Development of this new technology depends on providing readout on a much finer scale—to attach the electronics to detector elements by connection to the surface of the detector rather than the edge.

2. *Speed.* If a detector is sensitive for a certain period of time, it will respond to all the particles traversing it in that time. If interactions occur several times during the sensitive period, the detector sees a corresponding number of background particles as well as the particles of interest. At the LHC, collisions will take place every 25 ns. If a detector is sensitive for say 75 ns, it will see an average background from two spurious interactions. Consequently, there are advantages to increasing the speed of response of detectors.

Radiation Damage

When a detector or its associated electronics are subjected to high radiation levels, as represented by the passage through or interaction of large numbers of particles with the detector medium itself, there is some probability of a modification to the material of the detector or the electronics. This can lead to a reduction in the signal from the detector or even more dramatic malfunctions. There are many examples.

A straightforward example occurs in the case of plastic scintillators. For the light pulse to exit the scintillator and be detected efficiently, the plastic should be transparent. Exposure to sufficient radiation excites color centers, and many plastics become brown. They thus transmit less light, and if the condition becomes severe enough, they must ultimately be replaced.

Silicon detectors are also limited in their resistance to radiation. These detectors must be carefully designed and the background environment accurately modeled to ensure many years of efficient operations at facilities such as the Tevatron collider or the LHC.

Typically, two avenues may be pursued to mitigate such effects. On the one hand, the materials used can be improved, which is the current situation with respect to silicon devices. Research is being conducted to modify those characteristics of the material that make it sensitive to radiation. As with the original development of silicon detectors, this requires cooperation between the user and the manufacturer. Often the problem involves impurities in the material at the level of parts per million or less. At this level, few manufacturing processes are understood. Furthermore, the materials science of the detector is not entirely understood. The particle physicist then has to grapple with issues of solid-state physics.

The alternative path is to look for new materials. There was hope for a few years that the use of gallium arsenide instead of silicon would be a solution. In terms of neutron damage, gallium arsenide held promise. However, measurements were then made with charged particles, pions, and protons, and gallium arsenide was found to be less resistant than silicon. A possibility currently under investigation is the use of commercial diamond crystals. This also looks promising, but considerable development is required before a viable and affordable solution can be identified.

Opening New Possibilities

Significant R&D efforts are aimed at developing the detector technologies that will be required to support future experimentation at electron, proton, and/or muon facilities. This section has described a few areas in which the needs for future development are accompanied by a sense of the directions that are most fruitfully pursued. However, it should be emphasized that progress is required on a wide range of fronts and breakthroughs can occur any time a physicist is presented with some new possibility, perhaps developed outside the confines of elementary-particle physics, that acts as a stimulus. Research and development of detectors with no apparent immediate application has always been, and will continue to be, a necessary component of particle physics research.

7

❖

The Role of New Facilities

OVERVIEW

In Chapter 5, the progress likely to be made in experimental particle physics in the coming decade is considered. An enormous new energy range will be available, primarily with the start of operations at the Large Hadron Collider (LHC) in 2005. The LHC will be a superb instrument of discovery that supports an exciting program well into the next century. U.S. physicists, along with physicists from many other countries, are well positioned to play a leadership role in building this facility and exploring the new physics it will make available.

The tremendous enthusiasm throughout the field for the LHC derives from its power to discover new phenomena associated with breaking electroweak symmetry. Simple, elegant, and unequivocal arguments indicate that there will be clues to the origins of electroweak symmetry breaking in the energy range accessible at the LHC. It is also possible that the first hints of the origins of electroweak symmetry breaking will be seen before the LHC turns on either at the Tevatron or the Large Electron-Positron Collider (LEP). However, the physics reach of the LHC will be much greater than that of currently operating facilities, and there is every expectation that major discoveries will be uncovered with this collider. Tables 6.1 and 6.2 summarize the current and planned high-energy facilities that will define the high energy physics program for the next decade.

Physicists are confident that some of the central questions of the field discussed in Chapter 5 can be answered conclusively with the program of the next decade. However, as this chapter shows, even the enormous power of the LHC is not likely to provide a complete understanding of electroweak symmetry breaking or to completely unravel the origins of mass.

Because facilities that must operate at the highest energies may take two decades to plan, design, and construct, it is essential to try to anticipate now what will be learned from the LHC and other colliders by the end of the first decade of the next century and to understand what questions post-LHC facilities may have to explore. In discussing the technologies being developed now in order to address physics issues anticipated in 2010, it is important to consider both near-term technology (for a collider that might be built in the first decade of the next century) and long-term technology (for colliders that might not operate until 2020 or later). The LHC can be built today because the planning phase of the project, as well as much of the research and development work to solve technical problems and reduce costs, started more than a decade ago in the early 1980s. This chapter considers questions that will have to be answered even after the LHC program is mature. The accelerator technologies and possible collider facilities needed to address these questions are considered here. As described in Chapter 6, much of the necessary research and development is already in progress to develop the technologies that will define the colliders and the physics program of the future.

THE LANDSCAPE IN 2010

This section describes three scenarios for the physics that may have been discovered at the LHC at the end of its first 5 years of operation. By this time, experiments currently running or being built at LEP, the Tevatron, the Stanford Linear Accelerator Center (SLAC), and the Cornell Electron Storage Ring (CESR) (as described in the previous chapter) should be finished or near completion. Although one cannot predict the future with certainty, making projections is essential in planning for long-term evolution of the field. The scenarios chosen, at least from today's perspective, effectively bracket the possibilities of the state of our understanding in 2010.

The primary missions of the LHC and of the Toroidal LHC Apparatus (ATLAS) and Compact Muon Solenoid (CMS) detectors are to find evidence for the mechanism of electroweak symmetry breaking and to uncover and explore the origins of particle masses. The simplest model of how gauge bosons, leptons, and quarks acquire mass, and how electroweak symmetry is broken, includes a single neutral boson, the Higgs particle. This mechanism makes sense only if the mass of the Higgs particle is less than about 1,000 GeV (1 GeV = 10^9 electron volts). However, with this model the mystery remains as to why the mass of the Higgs boson should be so small compared with the grand unification scale discussed in Chapter 3. A possible alternative model is supersymmetry, which in its simplest version replaces a single Higgs boson with five bosons—three neutral and two charged. For supersymmetry to be the source of electroweak symmetry breaking, several of the many Higgs states in a supersymmetric model must have masses less than 1,000 GeV, possibly much lower. The lightest neutral Higgs boson should have a mass of less than 130 GeV. The current experimental lower bound is about half of this and will be extended by experiments at LEP soon. In addition, supersymmetry predicts many other new particles that are superpartners of all the particles discovered so far. Another possible model is technicolor, which postulates a new, very strong force and a number of high-mass particles. This case can be distinguished from supersymmetry both by the types of new particles seen and by the way they decay. It is also possible that a solution not yet thought of will be revealed at the LHC.

How the LHC and its predecessors, LEP II and the Tevatron, will help physicists distinguish between these three basic options—a single Higgs, supersymmetry, or new strong interactions—is considered below.

• **Scenario 1:** If the Higgs particle exists, the LHC experiments should see evidence for it. Suppose, as an example, that a new particle with a mass of around 110 GeV decaying into a two-photon final state is discovered. This particle would be a very strong candidate for a Higgs. However, verifying that this is a Higgs and establishing experimentally that this new particle has all the properties required of a Higgs boson, such as the correct couplings to gauge bosons and fermions, may be difficult. To verify that this new discovery is a Higgs state, one would like to study many of its possible decays. To rule out the possibility that this particle is not, for example, one of several expected in supersymmetry, one will want to rule out other Higgs states with masses less than 1 TeV (10^{12} eV) or so.

The LHC probably will be able to detect the Higgs in only a few final states, and the Tevatron (for lack of luminosity) and LEP II (for lack of energy) will not be able to detect the state at all if it has a mass greater than about 100 GeV. It also may be difficult at the LHC to conclusively eliminate the possible existence of other Higgs states up to a mass of 1 TeV.

• **Scenario 2:** If supersymmetry is the source of electroweak symmetry breaking, physicists will see experimental evidence for its existence at the LHC, if not earlier at LEP II or the Tevatron. Some, if not all, Higgs states should be observable at the LHC, and the experiments are likely to uncover evidence for many of the other particles associated with supersymmetry, such as the partners to gauge bosons, quarks, and leptons. A huge amount of tremendously exciting physics would pour forth from the LHC as experimenters untangled the many new particles produced. The challenge will be to verify that the new signatures are consistent with supersymmetry and to test the details of the theory. There are a number of supersymmetric models, and distinguishing among them will be difficult. The question is whether measurements at the LHC will be sufficient to ensure that supersymmetry is tested and that the patterns of masses and decay rates of the new states are understood. Determination of the masses of all of the new particles and the relationships between their decays will require a large number of independent measurements, not all of which would be possible at the LHC.

•   **Scenario 3:** If electroweak symmetry is broken by a previously undiscovered strong interaction such as technicolor, LHC experimenters will see indirect evidence for its existence in measurements of the scattering rate of gauge bosons. There may also be direct evidence since in some models of the new strong interaction, there are new particles that can be detected at the LHC and possibly even at the Tevatron. However, in some models, all of the new particles are of such high mass that LHC experiments cannot detect them directly as resonances, and the experiments to establish strong symmetry breaking will then be very difficult. In this scenario, experiments at the LHC will provide evidence for the existence of this new strong force but may not be able to provide much information about its structure. Great challenges will remain for both theorists and experimentalists.

In considering these scenarios, it is already clear that the tremendous energy range of the LHC will enable a great step forward in physicists' understanding of nature. In the field of high-energy physics, there is consensus that it is crucial to explore this mass range, which will point the way for new investigations to explore the TeV energy scale and address major questions of the field, such as the mass scale of Higgs boson(s), the mechanism for electroweak symmetry breaking, and the properties of supersymmetric particles. This is a primary objective for the long-term future of particle physics. Other parts of the high-energy program may well make important discoveries, but definitive tests of our understanding at the most fundamental level will come from the highest-energy colliders.

FUTURE COLLIDERS

Because of the long time scale to propose and build a new high-energy facility, physicists are already deeply involved in researching the new technologies for colliders that will extend their understanding of electroweak symmetry breaking and mass generation even beyond what will be provided by the LHC. The research and development for colliders of the twenty-first century has been under way for more than 10 years in anticipation of this need, and technologies have progressed sufficiently that in some cases one can begin to move to more concrete proposals for a new collider to follow the LHC. Coupled with the technical effort are strong efforts in cost optimization to make the next generation of colliders affordable, as well as efforts to ensure that the next accelerator facility will be a truly international endeavor.

Chapter 6 describes the three types of accelerators being considered by the international high-energy physics community: an electron-positron linear collider, a muon collider, and a very high energy hadron collider. The committee believes that a lepton collider with energy expandable to 1.5 TeV and sufficient luminosity would contribute essential information, complementary to that from the LHC, toward understanding the fundamental physics of electroweak symmetry breaking. The committee has concluded that it is appropriate at this time to intensify the international effort that will lead, in timely fashion, to a detailed design study for an affordable collider facility. Such a design study with an accompanying cost estimate could be ready early in the next decade. It is also the committee's belief that an even higher-energy, higher-luminosity lepton or hadron collider might be required to fully explore the experimentally challenging question of strong electroweak symmetry breaking, and R&D on both of these technologies should progress vigorously.

The Physics Need

Based on scenarios in the previous section, some of the physics issues that future colliders might have to address can be enumerated. A collider operating in the wake of the LHC will have to be capable of discovering some new particles and ruling out others. For example, if scenario 1 is correct, it will be imperative to verify that electroweak symmetry is broken by a single Higgs by ruling out any other additional Higgs states up to a mass of 1 TeV. This will require, among other things, a new collider of very high energy. Another desirable feature is the ability to explore more decay modes of the Higgs. If, instead, scenario 2 is the path nature has chosen, a new collider will have to clarify the details of supersymmetry that have not been explored at the LHC. In particular, some supersymmetric states are very difficult to detect at a proton collider but form a very important piece of the supersymmetry puzzle. As an example, in some models the supersymmetric partners of leptons will be undetectable at the LHC. Finally, a new collider would have to provide more details on the strong symmetry-breaking mechanism if a very high energy strong interaction is discovered at the LHC, as in scenario 3. This will require high energy and high luminosity.

Colliders to Address the Physics Need

Since it is not known which one (or more), if any, of the three scenarios discussed is correct, one must consider how a collider that might be operating as the LHC program reaches maturity would address all three possibilities. Here, features of the three types under development are reviewed briefly.

Various accelerator laboratories in the United States, Japan, and Europe have been carrying out R&D on construction of an electron-positron linear collider that would operate with an energy of more than 1 TeV. This is the most technically mature accelerator research program, with design concepts well advanced for facilities with energies of 1 TeV, luminosities (a measure of the number of beam particles passing through a given area per second) of 10^{34} cm^{-2} s^{-1}, and ideas for extending the energy to 1.5 TeV. In many instances, the necessary subsystem components required for a 1-TeV facility have undergone proof-of-principle demonstrations. Current challenges to design efforts include cost reduction, systems integration and operability, and the development of a cost-effective path to 1.5 TeV. Detailed studies have explored the physics reach of the electron-positron linear collider and how it might address the issues raised in the scenarios discussed here.

The concept of a muon collider is much more speculative than an electron-positron collider, but it has generated much excitement because potentially a muon collider could operate at much higher energies. Research on muon colliders has begun at Brookhaven, Fermilab, and Lawrence Berkeley National Laboratory. Discussions have focused on a collider with 4-TeV collision energy and 10^{35} cm^{-2} s^{-1} luminosity. The research and development for a muon collider is less advanced than that for an electron-positron collider. Since such technology has never been attempted, a low-energy prototype may well be needed as a demonstration before a full-scale collider can be proposed.

The challenge to the research program for very high energy hadron colliders is once again cost reduction, and research in cost-effective magnet designs is being pursued at Fermilab and Brookhaven. Detailed designs for an accelerator do not yet exist, and detailed studies exploring the physics potential of such a collider have just begun. Discussion is focused on a collider of 60-100 TeV of energy and a luminosity of 10^{34} to 10^{35} cm^{-2} s^{-1}.

The three scenarios can now be revisited to see how these new colliders could address the physics needs defined earlier.

• **Scenario 1 Revisited:** If a Higgs state of 110 GeV is discovered, the challenge to the LHC and a new collider will be to try to prove conclusively that this state is a single Higgs and not one of several such states expected from supersymmetry. Information to probe the Higgs couplings would be needed, and the possibility of other new Higgs states with masses up to about 1 TeV would have to be explored effectively.

Most events in a lepton collider come from direct annihilation of the lepton and antilepton, and these events are simpler and often easier to interpret than those produced at proton colliders. As a result, experimenters at a lepton collider may be able to isolate signals for several additional final states of the Higgs, such as Higgs decays to bottom and charm quarks and Higgs decays to tau leptons. This will allow experimenters to further probe the properties of this new state and test whether this Higgs couples to quarks, leptons, and gauge bosons with the expected strengths.

In the low-background environment of a lepton collider, often the discovery of new particles is relatively easy. If the collider operates above the energy threshold necessary to produce them, new particles can be directly detected. If the collider has sufficient luminosity, experimenters can search for indirect evidence of new particles to even higher masses. For example, detailed studies show that with an energy of 1.5 TeV in the center of mass and a luminosity of 10^{34} cm^{-2} s^{-1}, experiments at a lepton collider would directly or indirectly find evidence for any charged supersymmetric Higgs particles, if they existed, to a mass of well over 750 GeV and other neutral Higgs states to well over 1 TeV. Therefore, if only one Higgs state was seen, supersymmetry at the electroweak symmetry breaking scale could be ruled out, and scenario 1 would be established.

- **Scenario 2 Revisited:** The discovery of supersymmetry at the LHC, if not before, would be an important verification of some of the most exciting theoretical speculations of the last 20 years. To understand the dynamics of supersymmetry breaking and how the electroweak gauge symmetry is broken will necessitate information from both a proton collider such as the LHC and a lepton collider. Detailed studies show that it might be necessary for the lepton collider to have an energy of at least 1.5 TeV to cover the full range of the supersymmetry spectrum. The information from a hadron collider and a lepton collider is largely complementary. Experiments at proton colliders would measure the properties of the supersymmetric partners of quarks and gluons. Experiments at a lepton collider would make measurements of the supersymmetric partners of leptons and of the electroweak gauge bosons. All of these inputs will be crucial in order to understand the details of supersymmetry.

Lepton colliders also have the attractive feature that the lepton beam can be polarized, which allows experimenters to selectively suppress troublesome backgrounds. The polarization can also be used to turn on and turn off different production mechanisms for the supersymmetric particles and to single out subsets of reactions. The ultimate goal would be to combine information learned at the LHC with that learned from a lepton collider. This would give physicists the power to fully test the supersymmetric relations between the couplings of various particles and to use the determination of the parameters of superparticles to study the mechanisms of supersymmetry breaking that reach to the highest mass scales.

- **Scenario 3 Revisited:** The discovery of a new strong force at the LHC would present exciting new challenges to experimentalists and theorists alike. It will be of the utmost importance to learn more about the very high mass particles associated with this force. A 1.5-TeV lepton collider will not be able to produce particles of the new strong interaction directly if these particles have a mass greater than 750 GeV. However, with sufficient luminosity, such a collider could find indirect evidence for their existence in the process $e^+ e^- \rightarrow W^+ W^-$, even if they have masses as high as 4 TeV. These measurements will be essential to learn the basic parameters of this new force and begin to understand it. A detailed understanding of the strong symmetry-breaking scenario, however, will likely require colliders of even higher energies and higher luminosities. For example, a lepton collider with an energy of 4 TeV or a very high energy hadron collider could probe the structure of the new strong interactions and directly produce some of the resonances if they exist. Studies demonstrating that the signals for strong electroweak symmetry breaking would be detectable above background at such colliders are now under way.

The committee concludes that a lepton collider, if it could ultimately reach 1.5 TeV in energy and a luminosity of 10^{34} cm^{-2} s^{-1}, would contribute essential information complementary to that from the LHC toward understanding the fundamental physics of electroweak symmetry breaking. The committee also concludes that a very high energy hadron collider or a lepton collider of energy substantially higher than 1.5 TeV might be needed to explore further the experimentally challenging regime of strong electroweak symmetry breaking.

THE NEXT STEPS

The three accelerator technologies discussed in the previous section are at very different stages of development, and the next step to take is different in each case.

For the electron-positron linear collider, the next step in the process is an intensified international effort leading in a timely fashion to a complete design and cost estimate for such an accelerator. This design report should address engineering issues such as systems integration and operability, as well as identify a cost-effective path for achieving an ultimate energy of 1.5 TeV with a luminosity of 10^{34} cm^{-2} s^{-1}. This step could be completed early in the next century.

With the muon collider, a technological basis for the accelerator must be established. If the accelerator concept proves viable, it will be necessary to produce a detailed design, to estimate costs, and to discuss the feasibility of a demonstration accelerator at lower energy to explore the new technologies required.

Continued R&D on cost reductions is necessary for the very high energy hadron collider, along with an intensified effort elucidating the physics potential of such an accelerator. When these are further advanced, it would then be appropriate to begin a serious design effort. This will prepare the community for the possibility that when the LHC program is mature its experiments will indicate the need for a significant increase in collision energy.

The next high-energy collider after the LHC would clearly constitute a very significant fraction of the high-energy physics program in the United States and worldwide. This will be true whether this accelerator is built in the United States or abroad. It must be fully supported by the U.S. high-energy physics community. To reach such a consensus within the field, several steps will be necessary:

- The physics potential of a new collider must be compelling.
- The many design challenges of the accelerator technologies must be understood.
- The costs to build and operate the collider must be thoroughly understood and compatible with realistic funding scenarios.
- The role that such a new facility would have in the overall elementary-particle physics program must be fully evaluated.

Furthermore, the United States government must be committed to significant support for such a facility, no matter where it is built.

The committee believes that early in the next decade, even before initial operations of the LHC, it may be possible for the U.S. community, in conjunction with the high-energy physics communities in Europe and Asia, to decide to pursue a particular technology and commit to building the next major collider facility. It is possible that discoveries in the intervening years will clarify the physics of electroweak symmetry breaking to the extent that one or more of the scenarios discussed could be eliminated.

Finally, the scale of particle physics accelerators is such that this facility will require strong international backing. Countries should rightly compete for the prize of having such a premier scientific instrument, but the commitment to an international collider facility means that we have to recognize and plan for the unfortunate possibility of no forefront accelerator in this country in the early decades of the next century. The management of the field of high-energy physics, both within and outside the United States, must allow and encourage truly international cooperation for all phases in the design and construction of such a facility, independent of its location. The groundbreaking progress that has been made in structuring the participation of U.S. physicists on the LHC project is an important step toward this goal.

8

❖

Accelerator-Detector Technology and Benefits to Society

INTRODUCTION

From the early scattering experiments of Rutherford to the colliding beam experiments that produced the top quark with a mass almost as large as Rutherford's gold nucleus, particle beams have been the mainstay of elementary-particle physics. The tools described in Chapter 6 are among the most technically sophisticated in the world. Elementary-particle physicists are motivated to develop and use these tools by a deep desire to understand how the world works at its most basic and fundamental level. These tools can, however, and often do find application beyond the restricted realm of elementary-particle physics research. This chapter discusses a few examples of the application of elementary-particle physics technologies to the more general benefit of society in areas as diverse as biology, medicine, microelectronics, and national defense.

In elementary-particle physics as in most fields of science, advances in understanding are closely coupled to advances in technology. Many technical obstacles exist in the search for fundamental physics, and much of the creative effort of elementary-particle physicists, both experimental and theoretical, is devoted to overcoming these obstacles: Higher-energy accelerators are needed to cross thresholds for suspected new phenomena, and machines of higher luminosity open opportunities for the observation of rare and unexpected processes. Advances in accelerator technology must be accompanied by advances in detector technology as more complicated particle collisions are produced. Further advances in computing technology are necessary both to enable the processing of complex data samples and to allow comparison between various theoretical models and experimental results.

Experimenters have often made great advances when they have been able to borrow a technology developed for other uses and modify it to allow advances in particle physics. Similarly, a technology has often been developed specifically to address the needs of the elementary-particle physics community but then has been adapted to meet the needs of society outside particle physics. Although it is often evident where the technical barriers in elementary-particle physics (EPP) reside, it is much more difficult to predict where the breakthroughs will be and how they will come about. Experience has shown, however, that innovative new technologies or innovative uses for existing technologies will find surprising applications beyond those originally conceived by the developers.

THE MACHINE FRONTIER

The development of particle accelerators has led to new tools for basic research, medicine, and industry. Synchrotron radiation, the bane of elementary-particle physicists in their quest for ever-higher energy electron accelerators, is now used in cutting edge research in the materials sciences and in studies of biological systems. Accelerator-generated proton beams produce pulses of neutrons when they strike high-atomic-weight targets; these neutron sources play an important role in understanding the chemistry and physics of materials. Low-energy proton and pion beams, having served the needs of the elementary-particle physics community 40 years ago, are now used routinely in medical diagnostics and therapy. Cyclotrons, whose technology was first developed in the 1930s, now find medical application in the production of isotopes used for positron emission tomography (PET).

Industry employs particle beams for ion implantation in semiconductor devices, sterilization of materials, and x-ray lithography via synchrotron radiation. Using intense proton beams impinging on a target to produce neutrons can be a safer alternative to nuclear reactors. Proposed applications include the production of tritium, the destruction of plutonium and other high-level radioactive waste from nuclear weapons production and nuclear power plants, and even energy production by initiating fission in a subcritical reactor. National security is strengthened by current research into accelerator sources for explosives and contraband detection, neutron and proton radiography, and weapons effects simulations.

SYNCHROTRON RADIATION:
USING X-RAY LIGHT TO SEE THE WORLD IN ATOMIC DETAIL

As described in Chapter 6, electrons, when forced to travel in circles, lose energy through the mechanism known as synchrotron radiation. For electron energies of a few GeV (10^9 electron volts), this radiation is in the form of x rays. The need to replace the energy lost to radiation is a limiting factor in the design of high-energy circular electron machines. However, the x-ray radiation produced by electrons as they travel in their circular orbits can be used to probe the macroscopic structure of matter.

Circulating beams of electrons produce the most intense beams of x rays in the world. Early in the development of electron storage rings, facilities such as the Stanford Synchrotron Radiation Laboratory (SSRL) and the Cornell High Energy Synchrotron Source (CHESS) were built to take advantage of the synchrotron radiation produced as a by-product of storage ring operation for high-energy physics research. Now, newly constructed dedicated facilities at Argonne, Brookhaven, and Lawrence Berkeley National Laboratory use high-current stored electron beams to provide synchrotron radiation for many researchers studying a tremendous variety of problems in different areas of science, including geology, biology, chemistry, physics, materials, and environmental sciences.

The high energy of the x-ray photon gives it both its characteristic wavelength of about 1 angstrom (Å; 10^{-10} m) and the ability to probe the inside of dense matter. Since 1 to 2 Å is the typical distance between the atoms and molecules that compose our world, x rays have proven to be the most appropriate type of light for revealing the smallest details about our chemical and physical surroundings. Much of the work done at synchrotron facilities utilizes x rays to determine the structure and function of a material down to its finest details, usually by locating individual atoms. The small but intense x-ray beam is especially important for illuminating small specimens. Compared with a typical laboratory x-ray source, a millimeter-sized specimen might be exposed to from one to a hundred million times more photons at a synchrotron facility. As scientists study ever smaller specimens or smaller features of large specimens, the need for intense, collimated synchrotron radiation continues to grow.

The semiconductor industry, for example, has a developing need for synchrotron radiation. Because the sizes of wires and junctions on an integrated circuit are now smaller than the wavelength of visible light, standard microscopic methods are no longer sufficient for viewing or characterizing devices. Synchrotron radiation is used to study structures on silicon surfaces measuring one-millionth of a millimeter (Figure 8.1). From these studies, one gains understanding of a variety of processes that contribute to the production of a semiconductor device.

As a specific example, the use of x rays has helped refine several steps in the method of growing gallium nitride, a rare, wide-bandgap semiconducting material that produces blue light. At present, invention of the blue-light laser is a top priority in the international materials community because it will have an enormous technological impact. This impact stems from the fact that blue light has a shorter wavelength than the current generation of red-light–emitting diodes. Thus, a blue-light CD-ROM, for example, will be able to read and write smaller

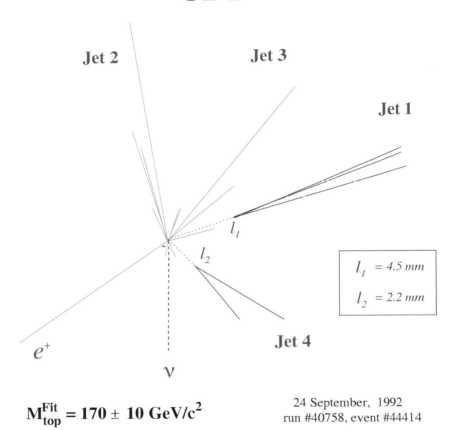

t̄t Event
SVX Display
CDF

Jet 2

Jet 3

Jet 1

l_1

l_2

l_1 = 4.5 mm

l_2 = 2.2 mm

Jet 4

e^+

ν

$M_{top}^{Fit} = 170 \pm 10 \ GeV/c^2$

24 September, 1992
run #40758, event #44414

FIGURE 4.6 Top quark event from the CDF experiment operating at the Tevatron collider. (Courtesy of the CDF collaboration at Fermilab.)

FIGURE 6.2 Picture of SLAC linac (site) looking upstream from downstream end. (Courtesy of SLAC.)

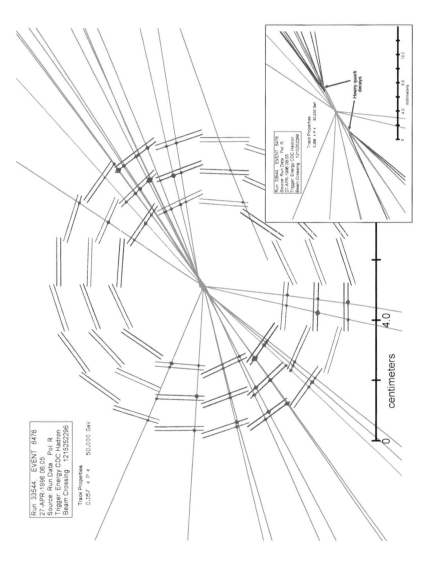

FIGURE 6.3 Picture of a reconstructed *B*-decay vertex from silicon detector of the SLAC Large Detector (SLD). (Courtesy of SLAC.)

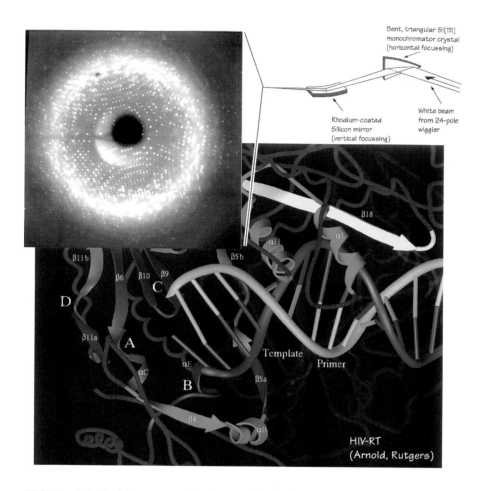

Labels within the image (top diagram): Bent, triangular Si(111) monochromator crystal (horizontal focussing); Rhodium-coated Silicon mirror (vertical focussing); White beam from 24-pole wiggler

Labels within the image (structural diagram): β18, αH, αI, β11b, β6, β10, β9, β5b, D, C, β11a, A, αE, Template, Primer, αC, B, β5a, β4, αB, HIV-RT (Arnold, Rutgers)

FIGURE 8.2 (*Top*) For macromolecular crystallography, x rays are focused onto a small sample to produce the diffraction pattern shown at left. (*Below*) Structural elements of the AIDS virus reverse transcriptase (RT) shown in relation to bound DNA. This type of picture, obtained from crystallography using CHESS, helps in understanding how polymerases can copy genetic material in living systems. (Proc. Natl. Acad. Sci. USA [1993] 90:6320-6324.)

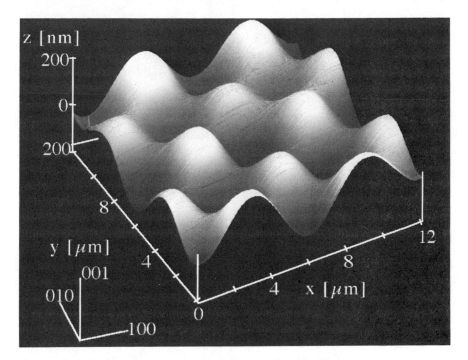

FIGURE 8.1 Atomic force microscope image of a periodic pillar structure fabricated onto the surface of a silicon crystal. Sample was previously subject to heat treatment, which relaxed the original square-wave structure into a smooth sinusoidal surface. (Courtesy of Jack Blakely and So Tanaka, Materials Science and Engineering Department, Cornell University.)

bits of information and thereby lead to an increased storage capacity of at least a factor of 4 relative to current technology.

In a different field of study, biologists have fully realized the efficacy of using synchrotron x rays to study large molecules of physiologically important materials such as proteins and viruses. Because these huge molecules have many thousands of atoms, a technique called macromolecular crystallography makes use of digital detectors and computers to collect and analyze rapidly vast quantities of x-ray diffraction data. Two examples of this research are the determination of the structure of the mammalian rhinovirus HRV14, which causes the common cold, and the determination of the structure of HIV reverse transcriptase (RT) type 1 (Figure 8.2, in color well following p. 112).

These research projects represent a subset of a larger effort on the part of the biological and biochemical communities to use x rays on a routine basis for determining the structure of living materials. Revealing their structure is just the first step in the process of understanding how proteins and viruses function. The most exciting and rewarding possibility then follows: By understanding the chemical structure and function of a virus, scientists hope to be able to design a drug that will inhibit the action of disease-causing agents. Although this field of so-called structure-based drug design is relatively new, it has already established itself as a nationwide (and worldwide) program, with a need for continuing access to synchrotron x-ray facilities.

One hundred years after the discovery of x rays, it is becoming evident that they can be more than just a tool to visualize atomic detail. The present trend toward ultraminiature design, exemplified by the rapid evolution of the integrated circuit, is creating the need to manufacture products with physical features smaller than the wavelength of visible light. A technique called x-ray lithography is just beginning to create micromechanical devices, machines with moving parts having dimensions as small or smaller than the width of a hair. It is easy to imagine that in the near future, x rays not only will play an increasingly important role in research and development but will have a significant impact on manufacturing as well.

SCIENCE AND INDUSTRY IN A PARTNERSHIP DOWN TO THE WIRE

Beginning in the late 1970s, Fermi National Accelerator Laboratory (FNAL) used thousands of spools of superconducting cable to build the magnets of the Tevatron, the world's first superconducting synchrotron. The use of superconducting magnet technology allowed a doubling of the magnetic field and a concurrent doubling of the energy. In the process of developing this technology, Fermilab brought experts in superconductivity together with physicists, engineers, materials scientists, and manufacturers of alloys, wire, and cable in a collaboration that helped boost the then-infant superconducting technology industry to a full-grown role in the billion-dollar world market created by magnetic resonance imaging.

Although "technology transfer" suggests the orderly flow of information from the laboratory bench to the factory to the marketplace, like all real-life human endeavors, technology transfer is more complicated and less orderly. In this instance, Fermilab's advanced research in high-energy physics required large quantities of superconducting wire and cable, creating overnight a market for products that did not yet exist. It was this "demand pull," rather than a particular new discovery, that launched technological development and large-scale production up a steep learning curve. Meeting this demand created an industry with the capability to supply a commercial market, driven by the important new medical diagnostic tool called magnetic resonance imaging (MRI), which had not been foreseen at the start of Fermilab's research.

To build the Tevatron, Fermilab used 135,000 pounds of niobium-titanium-based superconducting wire and cable between 1974 and 1983. At the project's start, the annual world production of these materials was a few hundred pounds. After Fermilab created the collaboration to develop large-scale manufacturing techniques, capacity grew so that today the annual production is in excess of 300,000 pounds, about half of which finds commercial application, principally for MRI.

The development of superconducting wire showed that science and industry could find common ground where both could thrive. Successful companies in the collaboration had the motivation to innovate, experiment, and invest resources to supply the materials to build the world's highest-energy particle accelerator, and in doing so, they built a new industry. Elementary-particle physicists did not invent MRI, but they did push superconducting technology out of the nest so that when MRI came along, the industry was ready to fly. The future can never be foretold completely, but indications are that superconducting technology has just begun to soar. In the words of the late Robert Marsh of Teledyne Wah Chang, still the world's largest supplier of superconducting alloys, "Every program in superconductivity that there is today owes itself in some measure to the fact that Fermilab built the Tevatron and it worked."

THE DETECTOR FRONTIER

The particle detectors utilized by elementary-particle physicists have evolved in sophistication over the past 50 years, providing the ability to investigate phenomena at continually advancing energy and intensity frontiers. Requirements for more precise spatial resolution, higher rate capability, and the ability to function in very high radiation environments stimulate the particle physics community to develop new and improved techniques. The evolution of detector technologies in support of elementary-particle physics has also led to advances in other fields such as nuclear and atomic physics and has found fertile areas of application in industry and medicine.

Tracking detectors date back to the invention of the cloud chamber in 1911 and the bubble chamber in the 1950s. Since these early years of elementary particle detectors, the tracking chamber has evolved considerably. The multiwire proportional chamber (MWPC) is a good example of a device that was invented and perfected for physics research and has had great impact outside the field. Georges Charpak (1992 Nobel laureate in physics for his detector inventions and their impact on science) invented the MWPC in 1968. The versatility of proportional chambers has led to many applications in medicine, materials science, and biochemistry.

Driven by the need to measure the energy and position of electrons and photons with very high precision, elementary-particle physicists pursued the development of bismuth germanium oxide (BGO) in a partnership with industry dating back to the early 1980s. During the past 10 years, the need for even higher spatial resolution and the desire to construct physically smaller detectors have led physicists to search for suitable alternative crystalline materials. A new development, lead tungstate crystals, is under way under the auspices of the Compact Muon Solenoid (CMS) experiment, which is under construction for the Large Hadron Collider project at CERN (the European Laboratory for Particle Physics). Industrial participation in this and other developments is not based solely on vendor expectation of an immediate financial payoff but also, and perhaps more importantly, on some form of secondary payoff. This could be from the potential for an extended market for the product or simply from the advantages of an enhanced reputation as a high-technology company.

The needs of several modern medical imaging techniques overlap the requirements of elementary-particle detectors described above in significant ways. A variety of techniques have been developed for nonsurgical imaging inside the body, such as computerized axial tomography (CAT), single photon emission computerized tomography (SPECT), and positron emission tomography. These and other medical diagnostic procedures depend on radiation detectors that have good spatial and energy resolution. Typically, short-lived radioactive isotopes are ingested by the patient in the form of drugs. The decay of these isotopes is then detected outside the body by detectors such as MWPCs. Detectors with sufficient spatial resolution can observe internal features of the body to a precision of several millimeters. Crystals, such as those described above, do an excellent job of measuring energy but cannot currently provide spatial resolution at the millimeter level, as required in medical applications, at an affordable cost.

An exciting new development in medical imaging is an outgrowth of the development of polarized gas targets for high-energy physics experiments. Based on a technique called "spin-exchanged optical pumping" targets of highly polarizable helium gas were developed in the 1990s for use in EPP experiments at the Stanford Linear Accelerator Center (SLAC) and at DESY in Hamburg. This technique has recently found application in a new MRI technique. In this case, the gas xenon is inhaled by a patient and, because of its enhanced polarizability relative to normal body tissue, an extremely high-resolution image of the gas-filled region can be obtained. Figure 8.3 shows a picture of a human lung, with a resolution far beyond that achievable by conventional MRI, produced using this technique. Other applications based on this new technology are being vigorously pursued in the medical community.

THE COMPUTING FRONTIER AND ELEMENTARY-PARTICLE PHYSICS

Elementary-particle physicists have made prompt and extensive use of the latest advances in computing for many decades and have contributed in important ways to the advancement of computing. Today, computing pervades the field with advances in detection, data acquisition, networking, and data analysis capabilities tied closely to the most advanced technical developments in the

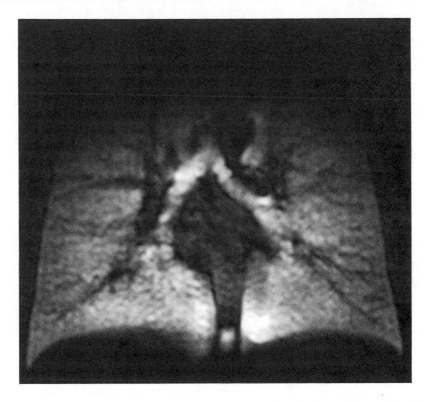

FIGURE 8.3 A picture of a human lung, with a resolution far beyond that achievable with conventional MRI, using spin-exchange optical pumping of polarized xenon gas. (Photograph courtesy of Gordon Cates, Princeton University; data courtesy of Duke University and Princeton University.)

computing field. Historically, elementary-particle physicists have created vast amounts of software for their research, with some having an impact on the world outside the field, for example, the invention of hypertext programming language and the World Wide Web.

The relationship between the elementary-particle physics community and the computing industry has evolved over the past 20 years. During the early years of the computer revolution, elementary-particle physicists were, by necessity, very creative in developing the computer applications and tools required to support their research. A prime example is the development of massively parallel processor "farms." To allow rapid analysis of events recorded in their experiments, physicists at particle physics labs started to build stripped-down (no monitor, no disk) computing engines. Vast arrays, or farms, of these computers were set to work on data reduction from experiments. Today, workstations and even

PCs are available in inexpensive configurations that are sufficiently powerful to serve as farms. This strategy, in one form or another, is being adopted (or rediscovered) anywhere that numerous similar computing tasks are required. Examples include statistical simulation (Monte Carlo) problems in other sciences, database searches in almost any field imaginable, and theoretical calculations as described in Chapter 9.

Other specific examples of the interaction between particle physics and the development of computing technology can be found in data acquisition for experiments. As small computers (minicomputers) were developed in the 1960s, the experimental particle-physics community recognized quickly their potential impact and developed computer-controlled data acquisition systems. The Computer Automated Measurement and Control (CAMAC) instrumentation standard was so developed. This consisted of a multicrate system of various modules, controlled from a master computer. The standard was very successful and found application in industry, which helped to advance its capabilities and usefulness. The variety of modules available escalated as manufacturers found a market outside EPP. Today, modern experiments challenge computing technologies with their data acquisition needs. For example, in a currently operating experiment known as KTeV, data are read into a distributed computer system for processing at the rate of 130 megabytes (MB) per second. Examination of these data for the purpose of making a real-time decision of which events to record permanently relies on a very large number of sophisticated computations requiring 30 processors of 200 MHz each operating in parallel.

Experiments in high-energy physics are now conducted by worldwide collaborations, requiring wide-area coordination of data handling and rapid communications. These social structures have made rapid use of the evolution of wide-area networks and effective new software standards. The need for effective networking facilities is particularly important to American scientists participating in experimental collaborations in Europe and Japan, but it clearly is an issue within the United States as well. It is, therefore, no surprise that physicists have been hooked up to the Internet and similar computer networks since these networks began. The general public, on the other hand, first began to appreciate the wonders of global computer communication in the mid-1990s. The ensuing popular explosion of the Internet can probably be attributed primarily to the development of a comfortable, universal hypertext interface for exchanging text, graphics, and images. The World Wide Web was invented at CERN to enable the organization and exchange of information between particle physics collaborators at locations all over the world.

As the information revolution picked up steam over the past 10 years, scientists were able to rely to a greater extent on industry to develop the computer tools needed to support elementary-particle physics research rather than having to develop these tools themselves. However, industry still finds the EPP community a valuable resource for testing innovations and future directions. The physics community provides an unquenchable demand for and knowledgeable evaluation of the technology. Computing technology will continue to advance rapidly, and elementary-particle physicists can be expected to apply the latest innovations to expand the reach of their discoveries.

TECHNOLOGIES FOR THE NEXT 20 YEARS

Continued advancement in the capabilities of the accelerators and detectors used in support of elementary-particle physics research is critically dependent on continuing advances in the underlying technologies. Although many of these advances are directly pursued in the laboratories and universities carrying out research in EPP, others occur independently as a result of research in other fields or in industry and are adopted by experimental particle physics. Likewise, developments in the technologies supporting experimental particle physics often find wider application in other areas of research or to the benefit of society at large. Although it is not possible to predict what technologies developed in support of experimental particle physics will impact the broader scientific research community and society in this country over the next 20 years, it is possible to indicate the technologies in which experimental particle physics is likely to invest considerable effort.

Superconductivity is the enabling technology for hadron accelerators and colliders. There is every reason to believe that the elementary-particle physics community will continue to pursue vigorously development in this field, often in close collaboration with industry. Recent rapid advances in the development of high-temperature superconductors give confidence that the development of a viable accelerator magnet, based on these superconductors, is likely in the first decade of the next century. Such a development could well parallel the first large-scale use of superconducting power distribution in this country and the enhanced availability of various superconducting-based medical devices. Although experimental particle physics will by no means be the sole driver for such a development, its participation will be significant.

Microwave power sources and accelerating structures represent the enabling technologies of electron colliders. Linear collider development can be expected to drive reductions in the fabrication and operating costs of such devices. Recent developments in periodic permanent magnet focusing klystrons could result in more affordable power sources and a corresponding increase in applications. Studies currently under way in the synchrotron radiation community are examining the possibility of utilizing the SLAC linac as a source of electrons for the world's first x-ray free-electron laser. Such a facility would provide significant new opportunities to the biomedical and materials research communities.

Detector development will continue to rely on increases in resolution and bandwidth. Pixel detectors will likely be required to support research at the very high luminosities contemplated at the Large Hadron Collider (LHC). Data transmission rates approaching 1 GB/s will be required. Distributed computing and extensive use of microprocessors will form the basis of both triggering and data analysis systems. Enhanced pattern recognition techniques and software algorithms will be required to sort out signatures in detectors recording tens of simultaneous events with hundreds of tracks each. What applications, if any, these developments will find in other fields is impossible to predict, but based on history it is safe to say that many will find application in areas beyond experimental particle physics.

❖

Interactions with and Connections to
Other Branches of Physics and Technology

INTRODUCTION

The questions addressed by elementary-particle physics have strong links to those of other disciplines. These connections are both intellectual and technological.

Today, different fields of physics appear to be becoming ever more specialized and this is largely true. Nevertheless, perhaps somewhat paradoxically, connections between fields are continuing to strengthen. This chapter indicates some of the strong areas of overlapping interest between elementary-particle physics and the disciplines of cosmology, astrophysics, nuclear physics, atomic physics, condensed-matter physics, fluid dynamics, and mathematical and computational physics.

COSMOLOGY

Unlike accelerator-based experiments, which are limited by available beam lines and interaction regions, the universe is an ever open laboratory. The possibilities for study, limited only by the imagination, are recognized by many particle physicists who turn to cosmology (and astrophysics, see below) for investigations of high-energy phenomena. Particle physicists with their expertise in large-scale computing and detector technology (low-noise and micropower analog and digital electronics, and large-scale and cost-effective detectors, to cite two examples) are well suited to contribute to such experiments probing the frontier of physics.

Cosmological observations have established that the known universe started as a small patch of space filled with radiation that subsequently expanded and cooled. In the earliest stages of its evolution, the universe contained extremely energetic particles, and it is likely that some important relics that could have been produced only at these early times are likely to remain today. Indeed, astronomical and astrophysical studies may very well be a way to study some aspects of particle physics beyond the Standard Model.

Particle physics interacts with cosmology on three important topics: dark matter, structure formation, and baryogenesis and nucleosynthesis.

Dark Matter

There is very strong evidence for a preponderance of dark matter in the universe, and there are strong arguments that it cannot be ordinary matter. This may be seen as one of the most important scientific discoveries of this century.

The amounts of various types of matter in the universe are customarily expressed as a fraction of what is called the critical density. The critical density is that value above which the universe will eventually contract and below which it will continue to expand forever. Of course the value of the actual total mass density in our universe is of great importance.

From observations, one gets important information on the dynamics of the matter in the universe. One looks at the radial dependency of rotational velocities of stars in spiral galaxies. These observations indicate that galaxies are made of visible stars in the center and are surrounded by massive halos of *invisible* matter. Additional observations using radio observation of the rotational velocities of neutral hydrogen in the gas clouds indicate that the halo extends way beyond the edges (defined by visible stars) of the galaxies. From these observations and also from studies of gravitationally bound clusters of galaxies, it is known that the mean mass density in the universe is at least 20% of the critical density and could be consistent with being equal to the critical density. In addition, there are indirect but compelling theoretical arguments as to why our universe actually has almost exactly the critical density; however, the question remains open.

Now, what is the composition of the matter in the universe? One can readily add up the matter in luminous mass (stars), and this is found to be less than 1% of critical density. Thus there is at least 20 times more mass in the universe that is nonluminous (i.e., "dark").

The question arises as to whether this matter is ordinary (baryonic) or possibly exotic (nonbaryonic). The amount of ordinary matter in the universe is, however, constrained by the big bang model to be between about 1% and 10% of the critical density. Thus, there is a major component of baryonic matter that cannot be accounted for in observations, and if the universe indeed has the critical density, then it is dominated by nonbaryonic dark matter. Even if not, there appears to be a significant component of the dark matter that is exotic. The prediction of baryonic density needs to be taken quite seriously, since the "big bang" model successfully accounts for the relative abundances of light nuclei and for the number of light neutrinos.

Two candidates for baryonic dark matter are hot, diffuse gas in galactic clusters and possibly in between galaxies and in their halos, and massive compact halo objects (MACHOs), such as brown dwarf, white dwarf, and neutron stars. There is strong evidence for both, and these two may very well make up the deficit in baryonic dark matter.

The nonbaryonic components are of primary interest to particle physicists. Elementary-particle physics provides unique insight into the nature of this component of dark matter. Candidates include neutrinos with mass, supersymmetric particles, axions, and magnetic monopoles. These candidates are primeval, nonbaryonic, and interact weakly with ordinary matter. They are treated here in turn.

Neutrinos

Neutrinos are one candidate for nonbaryonic dark matter. Unlike other candidates, they are known to exist; but to be a significant component of dark matter, neutrinos would have to have mass.

An upper limit to the mass density of the universe limits the sum of the masses of the three neutrino types: If the limit were violated, our universe would be contracting rather than expanding. Several experimental approaches are sensitive to nonzero neutrino masses. These include the study of neutrino oscillations (or mixing) in experiments at accelerators, measurements of the flux of electron neutrinos from our Sun relative to theoretical predictions, and studies of the ratios of neutrino flavors observed from the products of cosmic-ray interactions in Earth's atmosphere. The extensive experimentation in this area is discussed in Chapters 4 and 5.

Weakly Interacting Massive Particles

Another class of candidates for dark matter comprises supersymmetric and other weakly interacting massive particles (WIMP). The lightest supersymmetric particle is likely to be stable, so it becomes itself an interesting candidate for dark matter, particularly if it has a mass in the range from about 10 GeV (10^9 electron volts) to 1 TeV (10^{12} eV). In this range, WIMPs would be very slowly moving: They are *cold* dark matter candidates. They would be found clustered in our own galaxy and could be detected either directly (from interactions with laboratory equipment) or indirectly (from interactions in the halo of the galaxy).

WIMP signals are expected to be small, but detectors could take advantage of the unique signatures that WIMPs exhibit. For example, since WIMPs are orbiting in the galaxy, their velocity relative to the detector would exhibit predictable temporal variation. The rate of collision and the strength of the WIMP recoil would reflect this variation. The strength of the signal would also depend in a predictable way on the mass of the detector nucleus, the mass of the WIMP, and the relative velocity between them.

Three different types of detectors are under development for the detection of WIMPs. Solid-state detectors, consisting of large germanium and silicon ionization detectors, are used to detect WIMPs that collide with nuclei in the detector. The current limits on WIMP flux are set by these detectors. Sodium iodide and other high atomic mass scintillating materials show promise in increasing a detector's sensitivity to higher-mass WIMPs. WIMP collisions are also being sought in phonon detectors. The technology of detecting phonons (excitations in solids) is well developed in condensed-matter physics. When WIMPs collide with atoms in a crystal, such excitations are produced and the energy imparted to the crystal can be observed as a small rise in temperature, depending on the heat capacity of the material. At very low temperature, the heat capacity can be made very small, resulting in the best sensitivity, compared to other types of direct search detectors. Steady and significant progress is being made in the development of high-quality crystals and new temperature sensors.

Finally, indirect detection techniques are also being used. These involve searching for distinctive annihilation products of WIMPs. For example, WIMPs will be gravitationally captured by the Sun and annihilate within it, producing among other things high-energy neutrinos that can be detected in large underground detectors on Earth. WIMPs in the halo of our galaxy that annihilate can produce high-energy positrons and gamma rays or low-energy antiprotons that can be detected by instruments placed above the atmosphere. These indirect methods complement direct detection efforts in searching for particle dark matter within our galaxy.

Axions

Axion searches are discussed in Chapter 5; they exemplify the advantages of cross-disciplinary research. Axions are predicted to have mass, interact very weakly with matter, and are believed to have been produced at the time hadrons were produced in the big bang. Like WIMPs, as cold dark matter, axions would play an important role in galaxy formation, causing large-scale structures to evolve to what they appear to be at present and would be expected to cluster around our galaxy. The mass of the axion (if it exists) is now known to lie between one-thousandth and one-millionth of an electron volt: If it were any *lighter*, its density would comprise enough matter to be inconsistent with the age of the universe, whereas if it were *heavier*, some stars would shine for too short a time. Axions would be produced in the center of hot stars and act as cooling agents, speeding up stellar evolution.

Current searches are based on the fact that axions can be detected by their decay into two photons. The actual detection method uses a high-quality radio-frequency cavity to detect the presence of decay photons. Important advances in magnet and low-noise amplifier technology are expected spin-offs from such searches.

Magnetic Monopoles

A fourth candidate for nonbaryonic dark matter is the magnetic monopole. In string theories and in grand unified theories, magnetic monopoles are predicted to have been produced at the origin of the universe, but their density in the present universe must be low. They would be gravitationally bound to our galaxy and, if there are enough of them, would be observable. Currently, a large detector, the Monopole, Astrophysics, and Cosmic-ray Observatory (MACRO), has been built in the Gran Sasso tunnel in Italy to observe magnetic monopoles orbiting in this Galaxy. Within a few years of operation, MACRO should reach an interesting level of sensitivity.

Structure Formation

Astronomers have found a remarkable pattern of structure in the distribution of galaxies. Galaxies reside in large concentrations connected by thin, filamentary structures, surrounding large, quasi-spherical voids. These voids have typical scales of 200 million light-years. This structural complexity is contrasted with the smoothness of the very early universe observed in cosmic background radiation, where only tiny fluctuations of a part in 10^5 are seen across the entire horizon. So the question is, How did the universe evolve from such a smooth, featureless condition to the current one?

Cosmologists turn to particle physics for help in understanding the two basic issues underlying the formation and evolution of structure: the nature of dark matter and the origin and nature of the small density inhomogeneities that seeded all the structure. The unification of fundamental particles and forces are key here. As mentioned earlier, particle physics has provided several interesting possibilities for dark matter, and two attractive and very different possibilities for the origin of density inhomogeneities have been suggested. The first is that these inhomogeneities arose from quantum mechanical fluctuations during a very early, rapid period of expansion known as inflation; the second is that the seeds are topological defects, such as cosmic strings or textures, formed in a very early cosmological phase transition associated with breakdown of the symmetry between the fundamental forces.

At present, the most promising idea is that dark matter is slowly moving elementary particles (cold dark matter) and that the density inhomogeneities arose during inflation. However, both this idea and the possibility that seed perturbations were topological defects are being tested by fine angular scale measurements of the anisotropy of cosmic background radiation (the relic microwave radiation left after the big bang), the large-scale distribution of galaxies, and a host of other cosmological measurements. In studying the formation of structure, one is also exploring the unification of particles and forces of nature in a regime not accessible to terrestrial laboratories.

Baryogenesis and Nucleosynthesis

As discussed in Chapter 5, the universe appears to be made of matter, not antimatter (a small admixture of antimatter observed in cosmic rays can be understood as having been recently created through high-energy collisions with matter in the universe). To evolve from the initial condition of equal parts of matter and antimatter requires that the baryon number must be violated. The search for this violation is one motivation for experiments looking for decay of the proton.

Even more sensitive direct searches for antimatter are under way. Detectors on Compton Gamma-Ray Observatory (CGRO) are measuring spectra to search for characteristic gamma rays that would result from the annihilation radiation due to the interaction of an antimatter galaxy with a nearby matter galaxy. These searches, which can be done only in space with instruments having fine energy resolution, will be extended with new orbiting satellite gamma-ray experiments, such as Gamma-Ray Large Array Satellite Telescope (GLAST), under consideration by the National Aeronautics and Space Administration (NASA) and the Department of Energy (DOE). If there are antimatter galaxies, then cosmic rays of antimatter should also exist at a small level. The best limits on the antimatter component of extragalactic cosmic rays are now provided by the Balloon-Borne Experiment with Superconducting Solenoidal Spectrometer (BESS) detector; the Alpha Mass Spectrometer (AMS) detector is a new instrument being built to greatly extend the sensitivity. Should these detectors positively identify, for example, an antihelium component in cosmic rays, the implications for cosmology and particle physics would be highly significant.

There is an important interplay between particle physics and big bang nucleosynthesis. The formation of light nuclei such as deuterium, ^{3}He, ^{4}He, and ^{7}Li depends critically on the properties of neutrinos, such as the number of light neutrino flavors, and their mass and mixing parameters. This interface between the fields works in both directions, with the known neutrino properties providing constraints on the cosmological calculations and the astrophysical measurements (such as the abundances of the light nuclei) constraining the unknown particle physics parameters.

ASTROPHYSICS

Particle physics has contributed broadly to the issues of astrophysics. Three specific areas that have been directly impacted are the physics of the Sun, the physics of supernova explosions, and the study of very energetic cosmic rays.

Physics of the Sun

For more than 25 years, observations of the solar electron neutrinos have shown a significant deficit from the expectations of standard solar models. This observation has challenged the modeling of the Sun, motivating significant effort to see if modifications could explain the deficit. The theoretical understanding of the Sun has thus been significantly advanced, leading to the strong suspicion that the physics of the neutrino itself is likely responsible for the observations.

Supernovas

The theory of supernova explosions depends partly on the Standard Model. Neutral current interactions are needed to produce the explosion. In addition, an experimental contribution to the understanding of supernova explosions resulted from the serendipitous observation of neutrinos from Supernova 1987A by two proton-decay experiments, one in the United States and one in Japan. These experiments—both using massive, underground detectors—were built and operated to search for the decay of protons and were successful in establishing that the proton lifetime is greater than 10^{32} years (see Chapter 4). These experiments, being well shielded from most sources of cosmic rays, are also quite sensitive to neutrinos from astrophysical sources. They observed a pulse of neutrinos passing through Earth just before the arrival of visible light from the supernova SN1987A. It has been unambiguously established that these neutrinos came from Supernova 1987A. Measurements of the characteristic of these neutrinos contributed significantly to the theory and understanding of mechanisms acting within the extraordinary conditions inside a supernova. In fact, the neutrinos released by a supernova are a window right into the center of the "fireball," since once generated, they pass relatively uninhibited through the outer region of the exploding star directly to the detector on Earth. The visible light traditionally observed is that emitted from the surface of the "fireball."

Cosmic Rays

Cosmic rays are high-energy particles impinging on our atmosphere. How they are accelerated is a major unanswered question; they may also embody information about the early universe. Many of the astrophysicists who study the highest-energy cosmic rays were trained as elementary-particle physicists.

Cosmic rays represent the highest-energy particles ever detected by mankind. The most energetic (seen by the Fly's Eye detector in the United States and the AGASA detector in Japan) have 300 million times the energy of protons accelerated at the Fermilab Tevatron. Cosmic rays in this energy region almost certainly originate within the local supercluster of galaxies (those galaxies within a radius of about 60 million light-years from our Milky Way); otherwise they would be attenuated in their travels through the 3 K cosmic microwave background radiation left over from the big bang. The nature of the mechanism that accelerates them to these energies is completely unknown. Their energies represent the most significant departure from thermal equilibrium found in the universe. Some theorists speculate that they might be produced by exotic processes, for example, the collapse of massive cosmic strings, possible relics of the early universe.

It is clear that more data are needed to untangle this mystery. The Fly's Eye detector is now being upgraded to extend its sensitivity. The Telescope Array Project would view cosmic-ray initiated particle cascades in the atmosphere via their fluorescent glow. NASA is funding a 2-year feasibility study for a project to place an optical detector in Earth orbit that looks down and detects particle cascades in the atmosphere initiated by cosmic rays with energies greater than 10^{20} eV. One very ambitious proposal is the Pierre Auger Project. It has chosen a detector technology and two sites (one in Argentina and the other in Utah), both of which would be equipped with very large ground-based detectors. These and other new detectors will also study characteristics of high-energy collisions (in the atmosphere) in an effective energy region 200 times higher than that of the Tevatron. Although the interaction rate is much lower than at an accelerator, new phenomena could still be uncovered.

In addition to the protons and nuclei that enter Earth's atmosphere from the cosmos, there are gamma rays and neutrinos. Having no electric charge, these types of cosmic rays are thus undeflected by galactic and intergalactic magnetic fields. They point back to the very energetic astrophysical sources that produced them and tell astrophysicists about these sources. Very-high-energy gamma rays can be produced by synchrotron radiation and neutral pion decay. Neutrinos are produced largely in the decay of a charged pion.

Very-high-energy gamma rays are emitted by sources in our galaxy and by very energetic sources beyond. Observations have been made using a variety of techniques from Earth-based instruments around the world and from space. New observatories with much higher sensitivity are under construction. In some cases—for example, the Solar Tower Atmospheric Cerenkov Effect Experiment (STACEE)—these observatories use mirrors designed for solar energy research to focus light emitted in gamma-ray interactions with Earth's atmosphere into light sensors. A space mission to build a new observatory that will replace the aging high-energy gamma-ray instrument on CGRO is benefiting from the experience of elementary-particle physicists knowledge of detector design for measuring similar gamma rays in accelerator laboratories.

Observation of astrophysical neutrinos of very high energy requires extremely large detectors because of the very weak interaction of neutrinos with matter. Physicists and astrophysicists are instrumenting large volumes of water in lakes and oceans, as well as ice, to act as detectors for neutrinos. To obtain adequate sensitivity to very high energy neutrinos requires that a volume close to a cubic kilometer be instrumented. Initial studies by a Russian-German collaboration in Lake Baikal and by a U.S.-German-Swedish group using the ice at the South Pole have given promising results. There is a plan to extend the South Pole effort to the full size required for neutrino astronomy, and there are two efforts by European groups to instrument volumes in the Mediterranean Sea. An ocean-based observatory is also under discussion in the United States.

NUCLEAR PHYSICS

Fruitful interactions with nuclear physics have a long history and continue to the present. The discipline of elementary-particle physics grew out of the studies in the 1930s and 1940s of the atomic nucleus. Initially, techniques for particle acceleration were developed for the study of nuclei. The first working accelerators—electrostatic devices, cyclotrons, betatrons, and linear accelerators—were used for this purpose. Many techniques for particle detection, including proportional and ionization detectors, silicon detectors, sodium iodide (NaI) and germanium (Ge) detectors, and magnetic spectrometers, were initially developed for studies of nuclei.

Ongoing efforts in nuclear physics study fundamental processes and symmetries with nuclei and include solar neutrino studies and tritium beta-decay experiments sensitive to neutrino masses, studies of double beta decay, and tests of parity and time-reversal violation. These areas have greatly benefited from fruitful interactions between particle physics and nuclear physics.

Particle physicists study phenomena at the smallest distance scales, which require very high energy accelerators as tools. Today, the major focus of nuclear physics is the exploration of the structure of nuclei and single hadrons. To realize this goal, nuclear physicists for the first time are concerned with the quarks and gluons in the particle physics world as they live inside nuclear matter. The community has placed its highest priority on two facilities: the Continuous Electron Beam Accelerator Facility (CEBAF) and the Relativistic Heavy Ion Collider (RHIC), at the Brookhaven National Laboratory.

CEBAF promises to greatly expand understanding of the structure of nuclei and hadrons and to better constrain understanding of the consequences of the strong nuclear force; to improve understanding of the interaction between nucleons and strange baryons and the dynamics of strange baryons in nuclear matter;

and to study the origin of proton and neutron spin, a subject that has been the focus of a significant effort recently in both the nuclear and the elementary-particle physics communities.

RHIC is a good example of the scientific progress that can be generated when two fields (in this case nuclear physics and high-energy physics) pool their expertise in a synergistic way. The scientific question is one central to nuclear physics: Under what conditions do the low-energy constituents of nuclear matter (neutrons and protons) dissolve into quarks and gluons, thus decisively changing the nuclear many-body system. This question directly connects to two fundamental questions in physics: (1) What is the nature of quark-gluon confinement? (2) How did the early universe evolve from a quark-gluon plasma to the nuclear matter that makes up most of the visible mass of the universe today? To address such questions, nuclear physics has adopted some large-scale particle physics detection schemes, such as time projection chambers, further developing them for extremely high particle multiplicities.

ATOMIC PHYSICS

Many discoveries in elementary-particle physics come from experiments done at the highest possible energies. However, a great deal is also learned by very precise experiments at lower energies. From the high-energy physicist's point of view, atomic physics is the low-energy limit of the field. In atoms, one can study processes with extraordinary sensitivity. Because of the exquisite precision with which frequencies can be measured, these very low energy experiments effectively complement experiments done at higher energies.

There are several examples of particle physics done at atomic energies. One of the most rigorous tests of quantum electrodynamics comes from precision measurements of the Lamb shift, a slight shift in atomic energy levels due to fluctuations in the vacuum. Development of the experiments and the theory has now advanced to the point at which these agree at a precision of 1 part per billion. The neutral current weak interaction, which has been studied extensively at high-energy machines such as the Large Electron-Positron (LEP) collider and the Stanford Linear Collider (SLC), has small but observable effects in atoms. A combination of precision atomic measurements and new calculations of atomic structure has made possible precision tests of the Standard Model using the cesium atom. In fact cesium experiments provide nearly as stringent constraints on some non-Standard Model physics as do the precision experiments at LEP and SLC.

Another focus of particle physics done at atomic energies has been a search for time-reversal-violating forces. The observation of CP violation in K mesons leads naturally to predictions of T (time-reversal) violation, one consequence of which might be observable electric dipole moments for fundamental particles such as the electron or neutron. Recent improvements in laser and radio-frequency resonance techniques have led to experiments resulting in sensitive limits on these electric dipole moments. These limits can then be used to constrain models of CP violation and T violation.

One of the most fundamental assumptions of modern physics is that reality is invariant under charge conjugation, parity, and time-reversal symmetry (CPT) transformations. If so, then the absolute magnitudes of the masses, charges, magnetic moments, and mean lives of a particle and its antiparticle will be precisely the same. Measured properties of particles and antiparticles could differ despite CPT invariance if particles and antiparticles interacted differently with the apparatus made of particles alone (e.g., a long-range coupling depending on baryon number), although no such interaction has yet been observed.

CPT invariance must be subjected to rigorous experimental tests and has been, using the low-energy techniques of atomic physics. The magnetic moments of an electron and a positron have been compared to 2 parts per trillion. Low-energy techniques were used to make the measurement, and efforts are under way to increase the accuracy. Comparisons of the charge-to-mass ratios of a single antiproton and a single proton at an accuracy of several parts in 10 billion, by the TRAP collaboration, gave a CPT test at this accuracy. This collaboration also developed the techniques to accumulate cold antiprotons and positrons at 4 K for the production of cold antihydrogen and is now in pursuit of greatly improved CPT tests with leptons and baryons, which use laser spectroscopy to compare the properties of cold hydrogen and antihydrogen atoms.

CONDENSED-MATTER PHYSICS

Condensed-matter physics (CMP) and elementary-particle physics (EPP) share a deep conceptual unity. This is remarkable, given that these two fields operate on widely different distance scales. CMP deals with scales much larger than atoms, whereas EPP addresses scales one-thousandth the size of the proton or smaller. In many ways, however, there is a deep correspondence—both mathematically and physically—between phenomena in the two disciplines.

Understanding of the dynamical issues that occur in CMP (e.g., in superconductors) can be brought to bear on deep questions pertaining to dynamical arenas of EPP, such as quantum chromodynamics (QCD), understanding the electroweak physics scale of the Standard Model, and the Planck scale of gravity.

The cross-pollination of these fields has historically been a remarkable two-way street. For example, Richard Feynman first applied his path-integral techniques to quantum electrodynamics, the foremost particle physics problem of its day, then later to solve basic CMP problems, such as the behavior of superfluid liquid helium. Feynman diagram techniques are now universally applied in both CMP and EPP.

The intellectual parallels can be illustrated in the case of a superconductor where CMP asks the question: What is the structure of the state of lowest energy, whose properties determine much of the behavior of the system? The so-called Landau-Ginzburg model is a loose description, or a "toy model," of superconductivity. This model was superseded by a full and complete dynamical theory by Bardeen, Cooper, and Schrieffer (BCS). The BCS theory is one of the most remarkable dynamical models in physics. With it, conventional superconductors are understood.

Analogously, EPP, by attempting to understand the origin of quark, lepton, and gauge boson masses, is asking a very similar question: What is the structure of the vacuum, also the state of lowest energy, again whose properties determine why particles have masses and why weak forces become weak? (The vacuum in quantum mechanics is not nothing!) The vacuum pervading the entire universe can be thought of as a kind of superconductor, involving mechanisms that we are just now on the threshold of understanding. The Standard Model assumes that something like the Landau-Ginzburg toy model (slightly modified and redubbed the Higgs mechanism) is applicable. This gives a description of the mass generation of all quarks, leptons, and gauge bosons, and the rest of the machinery of the Standard Model performs beautifully in all experimental tests to date. Yet, the Higgs mechanism is really just a "black box" concealing a deeper mechanism that we do not yet understand, just as the Landau-Ginzburg model was a black box containing the BCS theory. Thus, EPP—with the Standard Model— finds itself today in a kind of "pre-BCS" era. The exciting aspect of all this is that we are on the threshold of understanding what is really happening by deeper examination of the physics currently accessible to Fermilab's Tevatron, and the LEP at the European Laboratory for Particle Physics (CERN) and eventually the LHC in the next decade.

Another remarkable connection between EPP and CMP is the study of topological structures that can occur in the vacuum, in complete analogy to "defects" that occur in solids when they form from cooling liquids. Indeed, entirely new branches of particle cosmology deal with the formation of these objects in the early universe and the problems and opportunities they create. As discussed earlier, cosmic strings or vortices are being actively considered by cosmologists as possible seeds for the formation of structure in the early universe.

Vortices and magnetic monopoles involve a profound connection between topology and quantum theory within modern gauge theories. Such topological objects are known to occur in some kinds of superconductors in CMP. They are understood in EPP to play a key role in the phenomenon of quark confinement in QCD. Other kinds of topological objects, called instantons, play a role in mass generation in the strong interactions and may arise eventually in our understanding of the weak interaction mass generation.

The techniques emerging from the abstract arena of superstrings in EPP are having an important impact in CMP. It is likely that the study of high-critical-temperature (high-T_c) superconductivity will lead to advances in EPP or that understanding of high-T_c will receive significant impetus from EPP results.

Other general mathematical methods emerging in EPP have had impact on CMP, and vice versa.

It is clear that in many ways, the true sister science of elementary-particle physics is condensed-matter physics.

FLUID DYNAMICS

High-energy physics experiments use special-purpose equipment and techniques, which can prove useful in other fields. For example, the large-scale production and storage of liquid helium at accelerator facilities requires designing and constructing specialized equipment that also finds application in the field of fluid dynamics. Fundamental experiments in turbulent flow at high Reynolds numbers require large-scale helium refrigeration equipment. The RHIC project at Brookhaven National Laboratory, which has the largest helium liquefier in the world, is a natural site for a large turbulent convection facility. Preparations are under way to build a Bénard cell 10 m high and 5 m in diameter operating with supercritical helium gas near 5.2 K and cooled by RHIC refrigeration.

Another project being considered in this field is building a wind tunnel large enough to test models of submarines that use liquid helium instead of air or water. The advantage is that a helium tunnel can reach operating Reynolds numbers to model nuclear submarines, something that cannot be done today. The large-scale cryogenic apparatus needed for such a tunnel already exists in large accelerator laboratories, and U.S. industry could fabricate such a device at the present state of the art.

Other spin-offs from high-energy physics to fluid dynamics are being considered. For example, some of the imaging techniques used in particle physics detectors could be adapted relatively easily to perform extremely fast tracking of particles seeded in a turbulent flow. Such an application would be a major boon to high Reynolds number research because the Lagrangian path of a seeded particle could be observed directly, something impossible to do today.

MATHEMATICAL AND COMPUTATIONAL PHYSICS

Research in physics has traditionally proceeded by two methodologies: experimental and theoretical. Traditionally, novel mathematical structures have been used heavily in constructing theories of particle physics, and now such structures are often invented by particle theorists. Both aspects of this exchange between particle theory and pure mathematics are especially evident today in string theory.

"Computational" physics is occasionally considered a third, comparably vital methodology. In reality, experimenters and theorists rely on computers to solve problems that would otherwise be intractable. The computing needs of "big" science have inspired many innovations. Often high-energy physicists have adapted ideas or technologies from other disciplines, developed them for their own needs, and returned a more powerful, practical product. One example is the effort in lattice gauge theory, illustrating the give and take with the computer industry and other branches of physics.

The most intensive computer jobs are in the domain of lattice gauge theory, part of the theory of elementary particles. As this work blossomed in the 1980s, it quickly became clear that the most powerful commercial supercomputers would be neither adequate nor cost-effective. A popular concept for reducing costs was to take commercial processors and connect them to each other. Such a computer is called "massively parallel" because very large numbers of processors compute simultaneously. With a massively parallel machine, one could, in principle, split up big problems and let each processor do a fraction of the job. The drawbacks are the difficulties of coordinating the split-up and of communicating the data among processors. Theoretical particle physicists decided to design and build parallel computers specifically for lattice gauge theory. They came up with elegant solutions to the coordination and communication problems, and the resulting machines are among the first practical examples of massively parallel computers. One of these consists of 8,000 50-MHz processors! Now, many computer vendors offer a parallel computing product.

The mathematical structure of lattice gauge theory has much in common with that of systems in condensed-matter theory, because both grapple with problems of large systems. After understanding the physical meaning of "renormalization" in elementary-particle theory, Ken Wilson sought a simpler problem on which to test his insights. He solved some outstanding problems of condensed-matter physics (later winning a Nobel Prize for the accomplishment) and came back from the experience with a way to define the theory of quarks and gluons on a lattice. Techniques of the resulting "lattice QCD" have developed side-by-side with condensed-matter theory ever since. In particular, computer programs running on massively parallel machines offer the most reliable way to work out details of the attraction between quarks inside the proton. As a result, many nuclear physicists have started to study QCD to understand nuclei as (complicated) composites of quarks and gluons.

10
❖
Elementary-Particle Physics in Today's Society

INTRODUCTION

Elementary-particle physics research, while being focused on the ultimate constituents of nature and their interactions and the technical aspects of their study, is embedded within society as a whole and has a sociological structure of its own. This chapter addresses several aspects of this side of elementary-particle physics. It includes discussions of how the field evolved, the role of universities and national laboratories, the effect of the demise of the Superconducting Super Collider (SSC), demographics and career advancement paths, governance, education, and public outreach.

Two other recently completed studies have addressed in some depth a variety of these issues: "Particle Physics—Perspectives and Opportunities" a report of the Division of Particles and Fields of the American Physical Society, and the report of the High Energy Physics Advisory Panel's subpanel on "Vision for the Future of High Energy Physics." Both serve as source material for parts of this chapter.

Recently, a comprehensive history of instrumentation, developments, and trends in the practice of research in particle physics has been published: *Image and Logic* (University of Chicago Press) by Peter Galison.

HISTORICAL BACKGROUND

Particle Physics Until World War II (the First 50 Years)

The search for the most fundamental building blocks of matter has been an important intellectual pursuit for most civilizations throughout history. J.J.

Thomson's discovery of the electron 100 years ago, using a simple particle accelerator (a cathode-ray tube), launched modern elementary-particle physics.

Until World War II, research in particle physics was usually carried out at universities, funded by university grants or by gifts and grants from corporations and wealthy individuals. Many early developments in the field originated in Europe. The invention of the cyclotron accelerator by E.O. Lawrence and M.S. Livingston in 1931 at Berkeley signaled the beginning of major U.S. participation in nuclear and particle physics. In the years leading up to the war, the United States. gradually achieved world leadership in these areas of physics, which were energized by the influx of physicists fleeing Europe and by the Manhattan Project, the nationally critical race with Germany to build the first bomb using nuclear fission.

It was during these years that the U.S. physics community began several major research and development programs funded by the federal government. The era of large science projects was born at laboratories such as Los Alamos and Oak Ridge, as well as at large nonnuclear facilities such as the MIT Radiation Laboratory. Projects were no longer accomplished by one or two senior collaborators assisted by graduate students and skilled technicians. Rather, a larger group of senior and junior physicists, together with professional engineers, developed and used large research facilities.

Particle Physics After World War II (the Second 50 Years)

Based on the success of controlling and using nuclear fission, a series of government agencies continued the wartime pattern of federal funding at U.S. universities. These have included the Office of Naval Research, the Atomic Energy Commission, the National Science Foundation (NSF), the Energy Research and Development Administration, and the Department of Energy (DOE). With the discovery of new particles such as pions and kaons in the late 1940s and the invention of new, more powerful accelerators, a dozen or so major universities built accelerators with energies above 100 MeV (1 MeV = 10^6 electron volts) to study new phenomena. The physicists who executed these projects applied the methodology of wartime laboratories: The machines were large, sophisticated engineering undertakings, compared to the tabletop experimental equipment of prewar research. During the 1950s, as the complexities of particle interactions and the rich spectra of meson and nucleon states began to unfold, elementary-particle physics diverged from nuclear physics and became a distinct field.

The need for higher particle beam energies required larger accelerators and correspondingly larger detectors and experimental facilities. Accompanying this increased size and complexity was an increase in the costs of construction and operation, eventually to an extent that outstripped the resources of a single university. In response, and again modeled on the wartime experience, large facilities were concentrated at federally funded laboratories. The national laboratories today are operated either by a single university or by a university consortium and support large user communities of university physicists. The efficiency of this form of cooperation between federally funded laboratories and the research community in universities played an important role in making the United States the undisputed world leader in particle physics through most of the past 50 years.

Four elementary-particle physics accelerator laboratories are operating in the United States today: Brookhaven National Laboratory, which has been in operation since 1947; and Fermilab, SLAC, and the Wilson Laboratory at Cornell University, which were all constructed over the 15-year period between 1957 and 1972. These facilities have been continually upgraded since their construction in support of the needs of the EPP research community. In 1983, a major facility authorized and under construction at Brookhaven called the Colliding Beam Accelerator (CBA) was terminated at the recommendation of the U.S. EPP community. Although the CBA would have added a new research tool, this recommendation followed the sense that the U.S. program could not do everything the high-energy physics community might have wanted it to do and that completing the CBA would have interfered with the Superconducting Super Collider. The existing facilities are currently engaged in upgrades that continue to afford scientists, not only from U.S. universities but also from around the world, the opportunity to do research at preeminent facilities. A funding history of EPP over the past 28 years is shown in Figure 10.1. The historic ebb and flow of funds between construction and operations of new accelerators is evident in this chart. Not included in this chart is the approximately $2 billion of expenditures on SSC construction from 1988 to 1993.

In Europe, particle physicists joined together in the middle 1950s to form the European Center for Nuclear Research (CERN) in Geneva, Switzerland. The organizational structure and administration of this pan-European laboratory was closely modeled after the U.S. laboratories. Over the past 40 years, CERN has established itself in friendly competition with U.S. laboratories as a major center for particle physics research and as a model for international organization. CERN's latest project, the Large Hadron Collider (LHC), will be the highest-energy accelerator in the world when completed. Germany, Italy, Japan, and Russia have also pursued active programs in experimental EPP over the last three decades.

Impact of the Termination of the Superconducting Super Collider

The Superconducting Super Collider (SSC) was to be the culmination of the rapid developments in particle physics research that occurred during the second half of the twentieth century. With 20 times the energy of the Fermilab Tevatron, a circumference of 53 miles, and a cost of about $10 billion, the SSC represented one of the largest scientific undertakings in the history of mankind. This premier

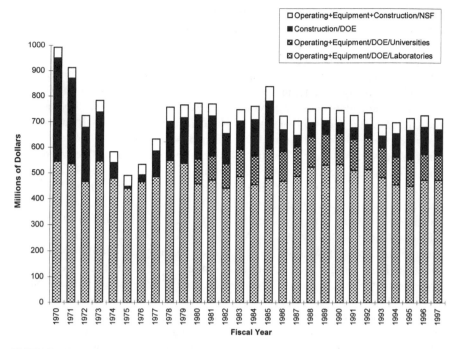

FIGURE 10.1 History of U.S. funding of elementary-particle physics (FY 1997 dollars) from 1970 to the present. Funding from both DOE and NSF is included in this graph. A total of approximately $2 billion of SSC construction funding between 1988 and 1993 is not included.

facility was designed to answer many of the fundamental questions described in this report and to search for unpredicted new phenomena. The SSC was envisioned by the U.S. elementary-particle physics community as the cornerstone of continued U.S. world leadership in particle physics into the twenty-first century. To many, both inside and outside this community, it represented a vital investment in our nation's ability to maintain its preeminence in basic research and to stimulate the development of new technologies in many areas that would contribute to the health of the U.S. economy. The demise of the SSC has had significant ramifications for elementary-particle physics in the United States and abroad.

Termination of the SSC in 1993 had a significant impact on the U.S. high-energy physics program, and this action still threatens U.S. leadership in the field. Its effect on the future vision of the U.S. program is still being sorted out, and the perception among some countries of the United States as an unreliable partner in international scientific endeavors was created. In addition, the high profile of the project, followed by abandonment of a significant investment, contributed to a view by a segment of the public of reduced U.S. commitment to basic research. The program at the existing laboratories and universities was further stressed by an accompanying erosion of financial support for operations, a stress partially ameliorated by significant upgrade projects that will keep the U.S. program in a leadership position through the middle of the next decade. However, when the LHC becomes operational in about 2005, the energy frontier will move to Europe. At present, there is no definitive plan to build a new forefront facility in this country.

The U.S. elementary-particle physics community has been redefining its future in the wake of the SSC cancellation and learning the lessons inherent in its demise. This report is part of that process. The demise of the SSC has been ascribed to many sources by many people. In any event it is clear that the ground rules governing the support of forefront basic research have changed. Among the lessons that might be drawn from the SSC experience are that a multibillion-dollar forefront facility can be undertaken only as a truly international project; that speaking clearly, frequently, and at the right level to the public on the research goals and opportunities in such projects is the duty of the physics community; and that multiyear funding and full authorization of projects are important to ensure stability. It is also clear that in the existing budgetary climate, every effort must be made to achieve significant cost reduction in base accelerator technologies. What has not changed, however, is the desire of the U.S. EPP community to pursue forefront research at the energy frontier, accompanied by the belief that this pursuit is necessary to retain a leadership position for the U.S. elementary-particle physics program. Strong and significant participation in the LHC program, accompanied by an ongoing, forward-looking, accelerator development program, are manifestations of these attitudes.

ORGANIZATIONAL STRUCTURES

Elementary-particle physics research is conducted as a partnership of universities, national laboratories, and governmental funding agencies. A recent survey of the field carried out by the Particle Data Group at Lawrence Berkeley Laboratory identifies 3,492 elementary-particle physicists (including graduate students) in the United States, distributed among 140 universities and 15 laboratories or institutes. The majority of researchers are based at universities. Universities also retain primary responsibility for the education of young people entering the field. National laboratories provide facilities and support for research beyond the resources of individual universities. Government funding agencies support this enterprise based on mandates given them by the public as represented by their elected officials.

Universities

University faculty members normally divide their time between teaching (i.e., helping students understand the accumulated knowledge in a field) and research (adding to the accumulation of knowledge). These activities are not disjoint but blend together in many important ways. Clearly the main function of graduate education in science is the training of future scientists, some of whom will continue to pursue basic science and increase our knowledge and understanding of nature. Others will choose to apply their scientific background and training to complementary pursuits.

Although graduate students are intimately involved in their university research programs, undergraduates also benefit from the research activities of their professors. It is typical for a university professor to have the assistance of an undergraduate working in his or her laboratory, contributing to scientific research, and getting trained in scientific techniques. Although most experimental particle physics occurs at sites remote from the university campus, undergraduate students can still participate. Many of the subdetectors of large experiments are researched, designed, built, or tested at university laboratories before being transported to the accelerator laboratory at which the experiment is to be run. Modern computer networks make it possible to analyze data that may be physically housed several thousand miles away. During the summer break, it is common for many undergraduate students to work at accelerator laboratories. All U.S. university groups pursuing elementary-particle physics research receive their funding either from DOE or NSF.

Laboratories

The four U.S. high-energy accelerator laboratories provide large accelerator and detector facilities to users from both U.S. and foreign universities. They are operated for the federal government by universities or associations of universities: Fermi National Accelerator Laboratory (FNAL) by the University Research Association (URA); the Laboratory of Nuclear Studies by Cornell University; SLAC by Stanford University; and Brookhaven National Laboratory by Brookhaven Science Associates (BSA). Astronomers and nuclear physicists have also emulated this structure by forming the Association of Universities for Research in Astronomy (AURA), which now operates several astronomical observatories as well as the Space Telescope Science Institute at the Johns Hopkins University, and by forming the Southeastern Universities Research Association (SURA) to operate the Thomas Jefferson National Accelerator Facility (formerly known as CEBAF).

DOE supports research activities at Brookhaven, Fermilab, and SLAC, whereas NSF supports the Cornell laboratory. University groups, whether funded by DOE or NSF, have free access to any of the laboratories. The work of these laboratories is guided and reviewed in a number of ways by the particle physics community and by funding agencies. Each laboratory has a visiting committee that reports to the university body that operates the laboratory. Funding agencies make periodic reviews of the physics research and technology development work of the laboratories.

At Brookhaven, Fermilab, and SLAC, external university users have formed user organizations. These work with laboratory administration on the problems of visiting physicists and graduate students, as well as other issues relevant to the research environment and capabilities of the laboratories.

Experimental Collaborations

Until about 20 years ago, elementary-particle physics experiments were pursued by relatively small collaborations. Experiments were usually designed to answer a specific question or a narrow range of questions. For example, an experiment might be proposed to measure a specific property of a particular type of particle. Laboratory beams would be prepared to produce this type of particle, and an experimental apparatus would be constructed to select and identify it and to measure the property in question. Data would be taken and analyzed, and the laboratory beams and the experimental apparatus would be reconfigured and modified for another experiment, which would often pursue different goals. This type of experiment was typically done by a single research group or by a few research groups collaborating, and many such experiments were run simultaneously.

Although there are still a few experiments of this type today, modern experimental particle physics is dominated by large collaborations, sometimes involving as many as 100 research groups and 500 or more physicists and students. Essentially every collaboration has significant membership from abroad. This situation reflects the emergence of colliding-beam experiments over the past two decades. As a result, detectors have become much larger and more complex, and the number of physicists needed to construct them has increased markedly. Typically, a research group, or collaboration of a small number of groups, constructs and maintains a specific piece of the detector. The modern colliding-beam detector serves in many ways the role of the traditional laboratory: It provides the data that individual research groups use to make specific measurements.

This analogy becomes particularly clear in the case of hadronic colliding beams, for example, the collisions of protons and antiprotons at Fermilab. The total number of interactions is very large, although the number of interesting events for a particular measurement may be quite limited. The experiments are limited in the number of interactions that can be recorded for analysis. Each experiment thus must select a mixture of events that are of particular interest. Typically, each experiment will have a committee to decide on what selection of triggers to use. Members of an individual research group might propose a particular trigger that would allow them to obtain a class of events necessary for the measurement they wish to make. They would then be allotted an appropriate fraction of the total data rate for their trigger.

This type of laboratory serving the needs of many research groups is common in other branches of science. For example, a large telescope will normally have a committee that allots viewing time to individual research groups to make particular observations in which they are interested. One significant difference between, for example, astronomers sharing a large telescope and particle physicists sharing a large colliding-beam detector is that it is the custom for all of the members of a particle physics collaboration to review the final physics results and sign the papers, even though the data analysis has usually been done by a smaller group of physicists. This practice recognizes the vital contribution of all the physicists who built and operated the experiment.

The Advisory System

The Department of Energy and the National Science Foundation hold the primary responsibility for directing an effective and well-targeted research program in EPP in the United States. Broadly speaking, these agencies have kept an excellent program in operation. For advice, they rely on several sources, the primary one of which is the High Energy Physics Advisory Panel (HEPAP). HEPAP was formed in 1967 as a standing committee to advise the administration on the issues it confronts in making decisions in particle physics. Its 15 members, named for 3-year terms, represent a broad cross section of university and laboratory physicists, both theoretical and experimental. The members are named by the Secretary of Energy, with the advice of the DOE director of the Office of High Energy Physics. HEPAP meets about five times a year. Its agenda is set by DOE and usually focuses on immediate questions faced by the department in particle physics, such as budget issues, program reviews, and international collaboration. HEPAP also appoints subpanels to study special questions on broad areas of planning. Its most important decisions relate to the overall direction of the field through its endorsement or rejection of proposals for construction of new facilities. The NSF Program Director for Elementary Particle Physics also regularly attends HEPAP meetings, and the NSF program is included within the purview of this panel. The successful pattern of HEPAP has now been adopted by nuclear physicists with the formation of the Nuclear Science Advisory Committee (NSAC).

The accelerator laboratories themselves have set up program advisory committees that review proposals for experiments and advise the laboratory director on the quality and feasibility of these proposals. The program committees have members from both the experimental and the theoretical side of the particle physics community with some members from abroad. Recently, the American Physical Society's (APS) Division of Particles and Fields (DPF), which includes most elementary-particle physicists in the United States, has become more active in policy and planning issues. The DPF, most recently in collaboration with the APS Division of Physics of Beams, organized a series of 3-week summer workshops in Snowmass, Colorado, on current questions of particle accelerators, detectors, and physics. These workshops provide an opportunity for the development of consensus within the EPP community and can often be used as a basis for long-range planning purposes.

Guidelines of the International Committee on Future Accelerators (ICFA) stipulate that large accelerator laboratories should be open to all qualified scientists, independent of citizenship. However, the selection of experiments and the priority accorded to them are the responsibility of the laboratory operating the regional facility. At the moment, for evaluating proposals at the international level, there is no advisory structure in place to make recommendations on the scientific merit, technical feasibility, or even the most appropriate accelerator laboratory for the execution of experimental proposals.

International Cooperation

Particle physics is a truly international undertaking. As in all basic science, research results are openly published and shared with other researchers around the world. Scientists from around the globe come to do research at our preeminent laboratories, and U.S. scientists have increasingly gone abroad to do research, most notably to CERN. Many of an entire generation of Japanese and European leaders of the field received at least a part of their training in the United States.

In the early days, international collaboration involved sharing primary data: Emulsions or bubble chamber films, exposed at U.S. accelerators, were shipped to scientists abroad for analysis. In the 1960s, as experiments continued to grow in size and complexity, their performance required collaboration by groups from many institutions. These collaborations frequently cut across national lines, the criterion for collaboration being the common interest in a problem rather than the common color of passports. As Europe, the Soviet Union, Canada, Japan, and China built their own accelerator laboratories, U.S. groups began to take advantage of unique opportunities abroad.

International collaborations were initiated, formed, and executed almost entirely by scientists working together to achieve a common goal. Sometimes they operated within a framework of bilateral national agreements, but on the whole, there were few government-level directives. Without a doubt, they have been remarkably successful. The sharing of talents and resources led to scientific productivity and improved cultural understanding. Scientists from nations that were enemies in World War II worked together immediately after the war. Even at the peak of the Cold War, productive experimental collaborations between scientists from the United States and the Soviet Union took place in both countries.

Future efforts to advance the high-energy frontier will almost certainly be pursued within a framework of multinational or international collaboration. Such collaborations will be necessary to advance both the hadron and the lepton frontiers. For the United States to be successful in this arena, new ground has to be broken in two areas: Collaboration must move beyond the realm of detector construction, data taking, and analysis, to that of design and construction of the accelerator facilities themselves. Because of the required scale of these efforts, an active role will have to be played by U.S. government agencies in negotiating the terms of participation.

An important step in this direction is being taken with negotiations for U.S. participation in CERN's LHC accelerator and detector construction. Participation in LHC is strongly supported by the U.S. particle physics community, recommended by HEPAP, and deemed economically feasible by the DOE HEP program on the basis of a constant-funding scenario over the period 1996-2004. A negotiating team from several government agencies including DOE, NSF, and the State Department has completed negotiations with CERN, and a framework for U.S. participation has been worked out. Participation is planned in the form of subsystem contributions, both to the accelerator (by DOE alone) and to the ATLAS and CMS detectors (by DOE and NSF), for which the United States would take full responsibility. The ability to develop such international agreements and then adhere to them over the multiyear construction period depends on congressional support and is critical for the success of any present or future collaborative effort.

Future Challenges

During the past decade, major changes in the structure and practices of the research program in elementary-particle physics have occurred. The rapid pace of these changes poses new challenges to the community as it strives to realize the full potential of the research effort with the most effective and efficient organization. This section reviews some of these changes briefly, along with the administrative and cultural problems that they pose. Many of these issues were studied by HEPAP's subpanel on Vision for the Future of High Energy Physics chaired by Professor Sidney D. Drell (the Drell panel)—the most recent HEPAP subpanel to review the field as a whole. A few passages of the commentary from this panel's report are quoted below. In addition, a recent survey conducted by Lawrence Berkeley Laboratory under the auspices of DOE, NSF, and the APS Division of Particles and Fields has developed a large amount of demographic information relating to trends in EPP. This information is accessible on the Internet at pdg.lbl.gov./doe_nsf/census.html.

Structural Changes in the Research Program

One of the most dramatic changes occurring over the past quarter of a century is the decrease in the number of high-energy physics laboratories and major experiments. Currently 55% of the U.S. high-energy physics budget is concentrated in two major laboratories, Fermilab and SLAC, which house all but one of the major domestic experiments in the U.S. program. This situation has been accompanied by a significant reduction in the number of experimental "spigots" at each laboratory over this period.

The evolution to larger experiments and fewer laboratories is a proper response to the demands of the science. These are the instruments needed to investigate the most important questions in particle physics, and the collaborations are of the size required to build and operate these instruments. However, it may be that the administrative structures inherited from a previous era are no longer optimal.

The Role of Universities

This change has had an impact on universities in that the engineering and technical capabilities of the field have become more centralized, away from university campuses. Management and construction of large facilities are more efficiently accomplished from the laboratory base. This situation has been accompanied by a real decrease in funding for the university component of the program. As shown in Figure 10.1, university funding has decreased by about 20% as measured in inflation-adjusted dollars over the decade of the 1990s. Many university groups are struggling to understand how to fulfill their traditional responsibilities in the newly evolved environment.

A significant part of the vitality and diversity of the field has come from faculty members' freedom to pursue the science that interests them without the constraint of working at a predetermined site. However, because of the concentration of resources on a relatively few large projects, this freedom has been somewhat reduced in recent years.

There is an additional concern about graduate education. The locus of graduate education, the training ground for the future generation of high-energy physicists, is at the universities that combine education and research. The difficulties of university groups in carrying out independent and viable research has severe impacts on the quality of graduate education, and there are concerns about the training of the next generation of scientists.

There is concern in the community that the present administrative structure does not fully reflect the concerns or address the problems of the university community. To quote the Drell panel report (p. 84):

> The university program does not have the same level of advocacy within the
> system as the national laboratories. National laboratories are represented by

strong directors who effectively advance the interests of their laboratories within the Department of Energy and are also capable of advancing their causes within the political arena. University groups, on the other hand, have no comparably visible advocate to represent their interest to the Department of Energy and to the National Science Foundation when they diverge from the interests of the national laboratories.

In a welcome response to these concerns, DOE has added three members to HEPAP who are explicitly charged with representing the views of university-based physicists working at accelerator facilities in the United States, working at accelerator facilities abroad, and conducting non-accelerator-based research in elementary-particle physics. In addition, the most recently convened HEPAP subpanel has been charged with, among other things, examining and making recommendations concerning the health and strength of the university component of the EPP program. Its report is expected in February 1998.

Internationalization

As discussed above, new large-scale experimental facilities will necessarily be international efforts. A likely scenario in the coming decades is one where a major portion of the high-energy physics research program will be concentrated at an international facility in which the United States is a major partner. This facility might or might not be located in the United States. More thought must be given to how such an international facility might be organized and administered and what impact participation in a large-scale international facility would have on the balance of the U.S. program. If the United States is to participate fully in an international facility in the future, new administrative frameworks will be necessary. It is important that these be carefully thought through as early as possible.

It is also true that elementary-particle physics projects now represent large and costly multiyear commitments. In recent history, such projects have been funded through the annual congressional appropriations process, without prior congressional authorization. This situation has hampered the efficient completion of some projects and, if it were to persist, would complicate U.S. participation in large-scale international projects.

Mechanisms for Setting Priorities

To carry out the best physics program possible with a fixed amount of funds, it is necessary to identify and pursue only the most promising opportunities. Pursuit of compelling new initiatives may require that some very interesting programs are either significantly delayed or never initiated and that others are terminated even while they are still productive. It is therefore extremely important that priorities are set in a way that optimizes scientific progress within the constraints imposed by available resources. The Drell panel report (p. 79) commented:

> Diversity, competition and alternative approaches to common scientific goals are necessary for maintaining the strength of the high energy physics program. However, it is wasteful to duplicate instruments and experiments without strong scientific and technical arguments. Plans must be carefully coordinated, particularly among national laboratories.

The setting of priorities in elementary-particle physics is done at two levels. At each accelerator laboratory, management receives advice from its program advisory committee and chooses the best projects and programs that fit within its expected budget. Then at the national level, the two funding agencies fashion a national program out of these components based on advice on priorities from HEPAP and related subpanels. Over the past 20 years, this process has left a number of very interesting projects unstarted or terminated because of hard decisions about priorities. Included are the CBA machine at Brookhaven, the tau-charm factory at SLAC, and a neutrino physics facility at Brookhaven. In addition, the start of a dedicated experiment for bottom quark physics at FNAL has been delayed. Ongoing programs that have been curtailed or reduced to make room for new construction include the Tevatron fixed-target program at FNAL, physics research at the Positron-Electron Project (PEP) ring at SLAC, and the activities of NSF university groups. It is fully expected that priority choices will continue to be required over the next few years, and additional program reductions will occur as new facilities begin operation.

Review of the Governance of the Field

There is some concern within the EPP community that existing administrative structures may not have kept pace with the above changes. In addition, it appears that EPP will face a number of critical issues in the coming years. As new information is acquired from ongoing experiments and progress is made in theoretical understanding, continuing adjustments to the overall objectives of the field will be required. Fashioning a viable and vital domestic program complementary to that at the LHC while contemplating the construction of a possible new facility will be an important challenge. The situation will likely depend on where the next facility is located. International collaboration and participation in new construction will continue to be critical issues for the field.

After reviewing many of the issues described in this section, the Drell panel report (p. 83) reached the following conclusion:

> Given these circumstances, the subpanel believes a thorough review of governance of the field is in order and should be undertaken by the supporting government agencies, the Department of Energy and the National Science

Foundation, in cooperation with the community through the American Physical Society's Division of Particles and Fields.

As yet, no such review of these very difficult issues has taken place, and it is probably still in order. There will be a need for continuing advice on these issues to Congress and the federal agencies that support research in this area, and a committee of the Board of Physics and Astronomy (which already has various standing committees in other areas) might be an appropriate avenue. An appropriately constituted committee would work with the agencies to undertake a comprehensive review of the system of administration of the research program in elementary-particle physics, to study the implications of increased internationalization of the field, and to explore possible alternative administrative and advisory mechanisms to respond to the changing environment.

EDUCATION IN ELEMENTARY-PARTICLE PHYSICS

When a country pursues excellence in particle physics research, there are many educational benefits. Early in life, children are attracted to science because it attempts to find answers to such fundamental questions as: What is the world made of? How did the universe begin? Will it ever end? Recognizing that answers to these and similar questions can be found through science, many are motivated to pursue studies in the physical sciences and mathematics. If seeking the answers to these questions is one of a nation's goals, and therefore socially encouraged, it provides a strong stimulus to scientifically oriented education. Not many of the children will end up studying particle physics, but they are more likely to obtain a solid grounding in science and mathematics that leaves them better prepared to cope with the modern technological world.

Particle Physics Graduate Education

Graduate students and postdoctoral fellows play a vital role in particle physics and also receive many educational benefits. They receive hands-on experience in the international high-tech environment of large centers and learn from direct contact with a staff experienced in many different technologies. They also learn to follow tight schedules and strict quality requirements, while facing stiff competition from other experiments. The annual number of graduating Ph.D.s in EPP has remained relatively constant at 160 per year over the past 25 years.

More than half of these students choose to pursue careers outside particle physics. They enter such diverse fields as the chemical or pharmaceutical industries, communications, computing and networking, the medical industry, investment banking, and the electronic components industry. They carry with them an ability to think logically, to solve problems, and to work effectively in collaboration with others. This continuing outflow of students into industry represents an efficient form of technology transfer.

Outreach to the Public

Research in basic science, of which elementary-particle physics is a major component, constitutes a substantial public investment in the foundations of our culture and in the well-being and technological advances of future generations. Far from being self-evident, the motivations, excitement, and significance of this endeavor must be communicated to the public. The important task of communication and outreach to the public can be achieved only in a collaborative effort among the media, schools, universities, and national laboratories. In the post-Cold War, post-SSC era, many elementary-particle physicists became convinced of the need to redouble efforts to communicate effectively to the public. This effort, which is a core responsibility of the scientific community, must go on continuously to build a base of understanding. In the past, communication with the public often occurred only sporadically (e.g., when support for a new, expensive project was needed). Very substantial efforts are already under way, and there exists a need to build on these existing activities.

An example of an attempt to foster greater understanding among the general public of the meaning and significance of scientific results is the "plain English" program recently initiated by the D0 experimental collaboration. As part of the normal publication process, collaborators explain new results and make clear how they fit into the picture of current particle physics research in language understandable to nonphysicists. These short summaries are currently available on the World Wide Web (http://www-d0.fnal.gov/public/pubs/d0_physics_ summaries.html), and the collaboration is looking into other methods of disseminating this information more widely. Thus far, the response, especially from science reporters, has been enthusiastic.

The national laboratories involved in particle physics research have also been influential in fostering science at the high school or college level. Most of these laboratories have programs aimed at bringing high school and college students into a research setting, putting them in contact with frontier researchers, and exposing them to real research environments. At some of these laboratories, such programs are geared toward groups, such as women or minorities, that have traditionally been underrepresented in science. In addition, the laboratories have special programs for high school teachers, which allow them to update their own training and then transfer that knowledge to students.

Both NSF and DOE have increased their efforts to create and organize materials about EPP designed to be intelligible to the public and useful to high school teachers. Some of these materials can be found on the Internet (http://www.nsf. gov:80/mps/phy/particle.html and http://www.er.doe.gov/production/henp/henp. html).

One very active national group is the Contemporary Physics Education Project (CPEP, http://pdg.lbl.gov/cpep.html), which consists of teachers, educators, and physicists. This group has created the wall chart on Fundamental Particles and Interactions and distributed more than 100,000 copies to U.S. secondary schools and colleges. It also has very popular color software for high school and college students. Packets of classroom activities about particle physics have been mailed to every high school physics teacher in the United States. CPEP conducts many workshops for teachers on how to use CPEP materials to teach particle physics.

A more complete listing of elementary-particle physics education and outreach activities at universities and national laboratories may be found at http://www-ed.fnal.gov/hep/home.html.

11

❖

Conclusions and Recommendations

INTRODUCTION

The field of elementary-particle physics has made dramatic progress over the past 25 years in understanding the fundamental structure of matter. Recent discoveries and technological advances are enabling high-energy physicists to address such compelling scientific issues as why elementary particles have mass, the excess of matter over antimatter in our universe, and the fundamental nature of the breaking of electroweak symmetry.

The United States is currently playing a crucial leadership role in the world-wide pursuit of the answers to these and other fundamental questions of particle physics. The accelerators and detectors at Fermilab, Stanford Linear Accelerator Center, Brookhaven National Laboratory, and Cornell are providing physicists from the United States and abroad with access to many experimental frontiers. In particular, the United States is home to the highest-energy accelerator in the world: the Tevatron at Fermilab. Furthermore, upgrades nearing completion at these facilities will ensure that the United States continues to be among the leaders in high-energy physics for well into the first decade of the next century. The elementary-particle physics program in the United States is complemented by unique facilities abroad that offer opportunities for U.S. scientists to do their research. In addition, an important part of the U.S. particle physics research program is performed without accelerators.

Leadership in elementary-particle physics research is dependent upon direct involvement in accelerator facilities operating at the high-energy frontier. In 2005, the energy frontier will move from the United States to Europe where the

Large Hadron Collider (LHC), with an energy seven times greater than the energy of the Tevatron, is now under construction. The LHC holds the promise of being a superb instrument of discovery, and there is every expectation that it will uncover important new phenomena. Maintenance of a forefront U.S. program in elementary-particle physics requires direct involvement in construction and utilization of this unique facility. The U.S. high-energy physics community is poised to play a leading role in both the construction of the LHC and its use in uncovering the physics of electroweak symmetry breaking. Although this physics is something completely new, strong theoretical arguments say that it must show up at the LHC.

The field of high-energy physics must now start to look beyond the horizon of the current program. As advised in Chapter 7, in order to explore the physics issues that are expected to remain open after the LHC, new accelerators and colliders will be needed. Given the long time scales for the design and construction of such facilities, it is essential to begin the planning process now. The ever-increasing size and cost of these facilities will require full international participation in all stages of their design, construction, and operation.

Maintenance of a leadership position in elementary-particle physics beyond the LHC era requires developing accelerator technologies to push the energy frontier ever higher, and it involves helping to build international consensus both on what technologies should be chosen for the next collider and on where the collider should be sited. Although the committee believes that it is highly desirable to have a forefront facility located within the United States, it is crucial that we maintain a technological base sufficient to allow full U.S. participation in all aspects of the design, construction, and operation of any such facility, independent of its ultimate location.

Over the past two decades, to address the important physics questions, very large experimental facilities have become the major instruments of the field. Accordingly, there have been changes in the way the field operates, particularly with respect to university groups. U.S. universities are recognized worldwide as being at the forefront of graduate education, and universities are the center of the process to educate future physicists. University groups, comprising approximately three-quarters of the experimental particle physics community, have always played a critical role in high-energy physics research. To understand how the challenging environment has affected university groups, the National Science Foundation (NSF) and the Department of Energy (DOE) have recently undertaken a comprehensive study of the situation. The committee applauds both agencies for taking this initiative.

In parallel with participation in large forefront facilities, it is essential to maintain an ability to support well-targeted areas of investigation, not necessarily at the energy frontier or accelerator based. Much of our knowledge of the field comes from experiments of great precision or great sensitivity at lower energies that take advantage of new detector or accelerator technologies. Within a reasonable budget, selected areas of investigation that show promise of having significant scientific impact must be supported.

It is now a fact that the time scale for construction and utilization of the primary research tools in experimental high-energy physics is well beyond a decade. This is much longer than could have been foreseen when the field began and when its management structure was set in place, and different mechanisms may be needed for its funding. Predictable funding is necessary for the effective management of research projects representing large and costly multiyear commitments. This is particularly so for international collaborations that the U.S. high-energy physics community may either join or host. In this regard, it should be pointed out that the earlier practice of full authorization at the start of major scientific projects led to efficient planning and construction.

RECOMMENDATIONS FOR U.S. ELEMENTARY-PARTICLE PHYSICS

The committee has developed its recommendations with two goals: (1) to exploit the great opportunities for discovery that lie ahead and (2) to maintain U.S. leadership in the field of elementary-particle physics. These goals require a diverse but focused program.

We are poised on the threshold of a new energy frontier, where discoveries are certain to be made, and new phenomena are likely to be revealed. This is the TeV (10^{12} electron volts) mass scale, where both well-established theory and revolutionary ideas predict new physics. First, the remarkable success of the Standard Model ensures that the secret of electroweak symmetry breaking will be revealed at this scale. Second, the exciting idea of supersymmetry, which offers the hope of great insights into unification of all the forces of nature, predicts that a rich array of new particles can be produced. Finally, we will obtain the first glimpse of physics well above the typical mass scale of the Standard Model. In the past, when such a large step has been taken, dramatic experimental surprises have occurred. One might expect that similar revolutionary discoveries will be made at the TeV mass scale.

The committee therefore believes that the highest priority is full involvement in TeV mass scale physics at large facilities uniquely suited to this purpose. This involvement includes exploiting the Fermilab Collider (presently the highest-energy facility extant); strong participation in construction of and research at the Large Hadron Collider (LHC) being built in Europe; and taking a leadership role in a future forefront international facility, possibly to be built in the United States. This path has historically provided the most fruitful avenue for uncovering new phenomena.

Other problems of great importance to the understanding of elementary particles do not require the highest energies for elucidation. One is understanding rare quark and lepton transitions. Another is the nature of CP violation—a phenomenon that bears on the apparent dominance of matter over antimatter in the universe. There are additional astrophysical questions of great importance that can likely be explained by particle physics dynamics, the most important being the nature of dark matter. A number of the most important findings in the field in the past two decades have been made by experiments studying problems such as these, and facilities presently being upgraded or under construction will allow such studies to continue. The committee believes it is crucial to support a well-targeted program in these areas. Given the limited resources that will be available, however, maintaining a proper balance between such efforts and those at the energy frontier will require difficult choices and keen foresight.

The committee's recommendations are therefore grouped into two classes: first, those relevant to the energy frontier, and second, one concerning important studies that are best done elsewhere. Both are essential to a balanced program.

Before presenting its recommendations concerning experimental initiatives, the committee comments on two subdisciplines of the field that are critical elements of a forefront program: non-facility-specific advanced accelerator R&D, which can lead to extension of the energy frontier, and theoretical physics, which provides the framework that organizes our observations.

Advances in elementary-particle physics have historically been tied to advances in accelerator technology. Accelerator research and development is of two general types—efforts targeted at the design and construction of specific facilities and more generic (and forward-looking) R&D targeting completely new methods of acceleration that will be required to support energy frontier facilities decades from now, should the physics demand it. This report contains specific recommendations with regard to the former. It is necessary to maintain an appropriate level of investigation in the latter area to secure the longer-term future of the field.

Theoretical work in elementary-particle physics provides the intellectual foundation that motivates and interconnects much of experimental research. The more formal areas of theoretical physics, especially string theory, hold the promise of providing a picture of the universe that accounts for an extremely broad range of observations and phenomena. The committee believes that a healthy level of activity both in formal areas and in the more phenomenological investigations that touch directly on experiments now and in the coming decade should be maintained.

1. Recommendations Concerning the High-Energy Frontier

At the present time, the Tevatron at Fermilab and the Large Electron-Positron collider (LEP II) in Geneva are the only machines operating at the energy frontier. In two years, LEP II will be dismantled, leaving the Tevatron alone at this frontier until completion of the LHC in the middle of the next decade. The LHC will dramatically extend the energy reach, pushing beyond the

TeV scale, where we know that the physics of electroweak symmetry breaking must appear. However, this report concludes that in the future, another collider will be required to complement or extend the range of the LHC and to explore fully the physics of the TeV scale. These considerations motivate a chronological structure for the committee's recommendations concerning the high-energy frontier.

1.a. Recommendation on the Fermilab Collider Facility

The United States should capitalize on the potential of the Fermilab Collider Facility while it has unique capabilities for investigations of high mass scale physics.

The Tevatron collider is the highest-energy accelerator in the world today and will remain so until the LHC era. The recent discovery of the top quark at this facility demonstrates its power to explore physics that is otherwise inaccessible. Its capabilities will be considerably enhanced with the new Main Injector. Although the LHC will be the first machine to extensively explore electroweak symmetry breaking, some of the new particles associated with the TeV scale might exist within the reach of the Tevatron. In particular, the upgraded Tevatron collider facility might discover supersymmetry. This would dramatically enhance our understanding of the universe.

1.b. Recommendation on the Large Hadron Collider

The committee enthusiastically endorses U.S. participation in the Large Hadron Collider project as a vital and essential component of the U.S. experimental particle physics program.

In the middle of the next decade, the LHC will supersede the Tevatron Collider as the highest-energy machine in the world. U.S. physicists, with their extensive experience at Fermilab and in the research and development toward construction and use of the Superconducting Super Collider (SSC), have established critical roles in the construction of the LHC machine and of the two largest experiments. The resources involved have been established in an agreement reached in 1997 by the Department of Energy, the National Science Foundation, and CERN (the European Laboratory for Particle Physics), the host laboratory.

The LHC will systematically explore a new energy regime, the TeV mass scale. LHC experiments will elucidate the mechanism of electroweak symmetry breaking, the central question of elementary-particle physics. The experiments will decisively test the prediction that a rich array of supersymmetric particles appears at this mass scale. If supersymmetry is indeed present at the TeV scale, the LHC will initiate the exploration of a vast new world.

The committee is convinced that participation in the enormously exciting physics promised by the LHC is essential for the vitality and continuity of U.S. particle physics. The committee also believes that U.S. participation is vital for the success of the project.

1.c. Recommendations on the Next Generation of Accelerators

As this report emphasizes, the committee anticipates that major discoveries will be made at the LHC. These will almost certainly point toward new phenomena that physicists will want to explore using an appropriate new collider.

Three types of machines have been discussed by the physics community: electron-positron linear colliders, muon colliders, and very large hadron colliders. Each has its unique capabilities and challenges, and each is at a different stage of development. Only the linear collider is far enough along to proceed to a conceptual design, with the engineering details and cost and schedule information appropriate to this stage. The other two options are sufficiently promising that increased research efforts are called for to make more realistic preliminary designs. These steps will put the community in the position to make a decision in the future about starting a new collider construction project with the best information possible.

A collider that complements or extends the reach of the LHC will require multiyear and multinational cooperation because of the magnitude of the resources needed. If the United States is to maintain a leadership role in this enterprise, it must participate both in accelerator technology development and in international decisions on the choice of technology and the location of the next facility. Although it is highly desirable to have a forefront facility located within the United States, it is crucial that the United States maintain a technological base sufficient to allow full participation in all aspects of the design, construction, and operation of such a facility, independent of its ultimate location.

Recognizing that it is too soon to endorse construction of any new machines, the committee makes recommendations concerning the development of each.

1.c.1. Recommendation on Electron Colliders

The committee recommends support of an international effort leading toward a complete design and cost estimate of an electron linear collider that would be able ultimately to reach a center-of-mass energy of 1.5 TeV and a luminosity of 10^{34} cm^{-2} s^{-1}.

An electron linear collider would contribute important measurements complementary to those from the LHC toward understanding the fundamental physics of the TeV mass scale. In the past, lepton colliders have been essential complements to hadron colliders. For example, W and Z bosons were discovered in a hadron collider, but many of their important properties could be determined only with the electron-positron colliders at LEP and the Stanford Linear Collider (SLC). For the physics of the TeV scale, this complementarity will likely continue to be important.

Laboratories in the United States, Japan, and Europe have been engaged for many years in research and development on an electron linear collider operating with an energy of 1 TeV or more. Stanford Linear Accelerator Center (SLAC), with its unique expertise in linear collider technology and the experience gained through the construction and operation of the SLC, is playing a critical leading role in these efforts. Many of the systems required for a second-generation linear collider have been or are being demonstrated. The next natural step is a complete design report, accompanied by cost optimization studies and a complete cost estimate.

The committee encourages the U.S. linear collider community to work cooperatively with international partners on the development of a common design and possible management structures.

The effort to complete an electron linear collider design and optimized cost estimate could be finished early in the next decade. It will then be necessary for the United States, together with the international elementary-particle physics community, to consider a number of factors in deciding whether to propose construction:

- The physics case for such a collider in light of any new discoveries in the intervening years;
- The construction and operating costs of the facility, together with the commitments and plans of the nations interested in hosting or participating in a linear collider; and
- The status of development of muon and hadron colliders.

1.c.2. Recommendation on Muon and Hadron Colliders

R&D targeted at developing the technologies for muon and very large hadron colliders should be vigorously pursued.

Experiments at the LHC may indicate new physics at energy scales significantly beyond those that it can decisively reach. Extension of the energy frontier beyond the reach afforded by the LHC will require the development of new technologies. A muon collider or a very large hadron collider has the potential for supporting even higher energies and luminosities in the post-LHC era. R&D efforts in both of these areas are in the early stages. Muon collider technology remains to be demonstrated, so the need is to focus on the development and validation of concepts. Present-day hadron collider technology could likely be used to construct a facility with a reach significantly beyond LHC, but the cost would be prohibitive. Hence efforts in this realm should focus on a reduction of cost through the use of advanced technologies. Development of both muon and hadron collider technologies must be pursued in a timely fashion to determine whether they represent technologically and economically viable options for reaching energies beyond those explored with the LHC.

2. Recommendation for Addressing Important Fundamental Physics Problems Below the TeV Mass Scale

The committee recommends strong support for a well-targeted program to study the fundamental particle physics that can best be explored with experiments below the TeV scale.

In its first recommendation, the committee has emphasized the range of important physics questions that are addressed at the TeV scale. It is important to recognize, however, that a number of outstanding fundamental questions can best be studied using other techniques. Foremost among these are the understanding of quark and lepton flavor mixing and of particle-antiparticle asymmetry (CP violation). There are also astrophysical questions of importance to particle physics, such as the nature of dark matter.

Experiments studying rare transitions between different families of quarks or leptons are extremely sensitive to new and interesting physics. For example, the 1964 experiment discovering CP violation found new fundamental physics that we are still trying to understand. One of the major themes of experimental particle physics in the next decade will be a systematic study of the interactions that mix the families of quarks and leptons.

Experiments in this area include several categories:

• Decays of the bottom quark. The central question to answer is whether CP violation is explained within the framework of the Standard Model or whether it is due to some new physics. The Standard Model explanation makes specific predictions that can be tested with very large samples of B mesons.

• Decays of the strange quark. Although CP violation was discovered in the decays of K mesons containing the strange quark, there are still outstanding issues in the CP-violating decays of strange particles. Experiments using extremely intense kaon beams give unique information about CP violation.

• Neutrino oscillations. Many experiments now give hints that a neutrino of one family can change into one of another family. One of the most important discoveries possible in the next decade would be unambiguous confirmation of any one of these hints.

A new era of research in these areas will begin in the next few years as experiments that should decisively answer many of the long-standing questions come on-line. Key U.S. facilities—the Positron-Electron Project (PEP-II) and the Cornell Electron Storage Ring (CESR) upgrade in addition to the Main Injector—will begin operations in the next few years with greatly enhanced capabilities to address this very important physics.

It is important to operate the newly built facilities and fund their critical experiments at the level required to take advantage of the physics opportunities they present. Historically, the U.S. high-energy physics community has phased out programs to accommodate those that are more scientifically desirable, and it should continue to do so. Because of limitations in resources for the field worldwide, in the future only initiatives that have the most promise for scientific advancement should be undertaken.

CONCLUSION

The recommendations above, if adopted, should maintain U.S. leadership in the field of elementary-particle physics well into the next century. They will allow our scientists to participate in what are likely to be profound and exciting discoveries, discoveries of a nature not seen before.

Appendix
Glossary, Abbreviations, and Acronyms

Å. See Angstrom.

Accelerator. A machine that increases the kinetic energy of charged particles such as electrons and protons for collisions with another beam or with a fixed target.

AGASA. Akeno (Japan) Giant Air Shower Array, a cosmic-ray detector.

AGS. Alternating Gradient Synchrotron, a 33-GeV proton accelerator at Brookhaven National Laboratory.

Amanda. Antarctic Muon and Neutrino Detector Array, a high-energy neutrino detector in the ice cap at the South Pole.

AMS. Alpha Mass Spectrometer experiment, a superconducting-magnet particle detector to be flown on the space station to search for cosmic rays of antimatter and dark matter.

Angstrom. A unit of distance, 10^{-10} m, denoted by Å.

Annihilation. See Antiparticle.

Antimatter. Matter composed of antiparticles (e.g., antiprotons, antineutrons, antielectrons) instead of particles (e.g., protons, neutrons, electrons).

Antiparticle. Each particle has a partner, called an antiparticle, with identical properties except that its electric charge and a few other properties are opposite those of the particle. When a particle and its antiparticle meet, they can annihilate each other.

Antiproton. Antiparticle partner of the proton.

APS. The American Physical Society.

Astrophysics. Physics of astronomical phenomena, such as the evolution of stars and the formation of galaxies.

Asymptotic freedom. Property of the strong force between quarks becoming weaker as quarks get closer together or as the energy of a collision between them increases.

ATLAS. A Toroidal LHC Apparatus, a detector being built at CERN to study proton-proton interactions at the LHC.

Atom. Smallest unit of a chemical element, approximately 10^{-8} cm in size, consisting of a nucleus surrounded by electrons.

Auger, or Pierre Auger, Project. A proposed cosmic-ray experiment made up of a large array of photodetectors. (See Fly's Eye.)

AURA. Association of Universities for Research in Astronomy, Inc.

Axion. A hypothetical low-mass boson and candidate for a dark matter particle.

B meson. Meson that contains one *b* quark and one *u*, *d*, or *s* antiquark.

B factory. Specialized accelerator facility that produces large numbers of *B* mesons.

Baryon. Type of hadron. The baryon family includes protons, neutrons, and other particles whose eventual decay products include the proton. Baryons are composed of combinations of three quarks.

BCS. Symmetry-breaking theory of superconductivity, for which John Bardeen, Leon Cooper, and John Schrieffer won the Nobel Prize in 1972.

Beam. Narrow stream of particles produced by an accelerator.

Beauty. See Bottom.

BEPC. Circular electron-positron collider with center-of-mass energy up to 6 GeV and high luminosity, located near Beijing, China.

Beta decay. Decay of a hadron by emission of an electron or positron and a neutrino through the weak interaction.

Bevatron. Circular accelerator at Lawrence Berkeley National Laboratory, Berkeley, California; previously used to accelerate protons up to 6 GeV, now part of a complex for accelerating nuclei.

Big bang. Standard Model of the origin of the universe involving an initial phase of high density and temperature followed by a expansion of space-time and cooling.

BNL. Brookhaven National Laboratory.

Boson. Particle with spin of zero or an integer value. Unlike the fermions, more than one boson can occupy the same quantum state.

Bottom. Fifth type of quark, also called the *b* quark or beauty quark.

Broken symmetry. A symmetry principle that is imperfectly respected.

Bubble chamber. Particle detector in which paths of charged particles are revealed by a trail of bubbles produced by the particles as they traverse a superheated liquid. Hydrogen, deuterium, helium. neon, propane, and freon liquids have been used for this purpose.

Calorimeter. Particle detector in which the energy carried by a particle or group of particles is measured.

CAMAC. Computer Automated Measurement and Control, a standardized electronic connection system, invented by nuclear physicists.

CAT. Computerized axial tomography, a means of imaging internal structures of objects using beams of x rays that probe different parts of the object from different angles.

CDF. Collider Detector at Fermilab, a detector experiment at the Tevatron.

CEBAF. Continuous Electron Beam Accelerator Facility, renamed the Thomas Jefferson National Accelerator Facility, located in Virginia.

Center-of-mass energy. The effective energy in a particle collision.

Cerenkov counter. Detector of Cerenkov radiation, which is electromagnetic radiation (usually visible light) emitted by a charged particle when it passes through matter at a velocity exceeding that of light in the material.

CERN. European Laboratory for Particle Physics (originally the European Center for Nuclear Research), located near Geneva, Switzerland.

CESR. Cornell Electron Storage Ring, an electron-positron collider with a center-of-mass energy of 10 GeV located in the Laboratory of Nuclear Studies at Cornell University.

CGRO. Compton Gamma-Ray Observatory, a multiple-detector NASA satellite.

Charm. Fourth type of quark, also called the *c* quark.

Charmonium. The family of hadronic particles composed of a charm quark and an anticharm quark.

CHESS. Cornell High-Energy Synchrotron Source, a national facility for energetic x-ray beams.

Circular accelerator. Accelerator in which particles move around a circle many times, being accelerated further in each revolution.

CMBR. Cosmic microwave background radiation from the big bang.

CMP. Condensed-matter physics, the study of matter in liquid and solid phases.

CMS. Compact Muon Solenoid, one of two general-purpose detector experiments (see ATLAS) at the LHC.

Collider. Accelerator in which beams of particles are directed to meet each other head-on to produce collisions.

Colliding-beam accelerator. See Collider.

Collimate. To produce a spatially narrow beam of particles or light.

Color. In high-energy physics, the property of quarks and gluons analogous to electric charge that determines how the strong force acts between a quark and a gluon.

Conservation law. Physical law stating that the total value of some quantity or property cannot be changed. For example, the conservation of energy states that the total energy of a system cannot change; this implies that the energy of all particles going into a reaction, including the energy associated with their masses, must equal that of all particles escaping from the interaction.

Cooling. In high-energy physics, a technique for decreasing the kinetic energy of a beam of particles, in the component of motion transverse to the beam direction.

Cosmic rays. Energetic particles that come from outside Earth's atmosphere.

Cosmology. A subdiscipline of astrophysics and astronomy having to do with the large-scale behavior of the universe and with its origin and evolution.

CP symmetry. Operation of changing a particle to an antiparticle (C) and left to right (P).

CP violation. Experimentally discovered phenomenon, in which CP symmetry does not hold.

Cryogenics. Science and technology of producing and utilizing very low temperatures.

Cyclotron. Circular accelerator design, made of two D-shaped magnets.

D meson. Meson containing a *c* quark and a *u*, *d*, or *s* antiquark.

D0. Detector experiment at the Tevatron at Fermilab.

DAPHNE. Electron-positron collider at Frascati, Italy.

DESY. Deutsches Elektronen-Synchrotron laboratory in Hamburg, Germany.

Deuterium. Heavy hydrogen, whose nucleus contains one proton and one neutron.

DOE. Department of Energy.

DORIS. Electron-positron collider that used to operate at DESY.

Down. One of the two lightest quarks, also called *d* quark. The other light quark is the up or *u* quark.

DPB. Division of Physics of Beams of the American Physical Society.

DPF. Division of Particles and Fields of the American Physical Society.

Drift chamber. Particle detector in which the passage of charged particles produces tracks of ionized gas. Electrical signals from these tracks are detected and recorded, allowing reconstruction of the particle paths.

Elastic collision. In high-energy physics, a particle interaction in which interacting particles are not changed into other particles.

Electric dipole moment. One possible way to distribute the electric charge of a particle that would violate basic symmetries.

Electromagnetic force or *interaction.* Long-range force and interaction associated with electric and magnetic properties of particles. This force is intermediate in strength between the weak and strong forces. The carrier of the electromagnetic force is the photon.

Electron. Elementary particle with a unit negative electrical charge and a mass about 1/1840 that of the proton. Electrons surround an atom's positively charged nucleus and determine the atom's chemical properties. An electron is a lepton.

Electron volt. Energy of motion acquired by a charged particle accelerated by an electric potential of 1 V.

Electroweak force or interaction. Force or interaction that unifies the electromagnetic force and the weak force.

Electroweak symmetry breaking. Mechanism that gives mass to W and Z particles.

Elementary-particle physics. Field of physics whose goal is to discover and understand the basic constituents of matter and the forces that act on them.

EPP. See elementary-particle physics.

eV. See electron volt.

Family. See generation.

Fermilab. See FNAL.

Fermion. Particles having the property that only one can occupy a quantum state (the Pauli exclusion principle). Such particles have half-integer values of spin.

Field theory. Theory that describes forces as originating from "fields" that permeate space. For example, electrostatic forces on an electron are due to the presence of electric fields that act on the charge of the electron.

Fixed-target experiment. Experiment in which a beam of particles is directed onto a nonmoving target. (See Collider.)

Flavor. Term used in high-energy physics; the uniquely characteristic "flavors" of the six types of quarks are called up, down, strange, charm, bottom, and top.

Fly's Eye. Giant air-shower detector array, located in Utah, exploring cosmic-ray events at energies greater than 10^{17} eV.

FNAL. Fermi National Accelerator Laboratory in Illinois.

G. Gauss, unit of magnetic flux density, equal to 10^{-4} T.

Gamma ray. Term used for photons with energies higher than the MeV range.

Gauge bosons. Particles, such as gluons, that have integer spin and carry force and are a manifestation of a symmetry.

Gauge theory. Theory of particle interactions, modeled on the immensely successful modern theory of electromagnetism, having a special kind of symmetry known as gauge invariance.

Geiger-Müller tube. Type of particle detector that works by sensing the ionization caused when a charged particle passes through it.

General relativity. Einstein's theory of gravitation and acceleration, which describes how the presence of matter or energy alters the geometry of space and time.

Generation. Classification of leptons and quarks into families, according to their internal properties. The first generation consists of the electron and its neutrino and of up and down quarks. The second generation consists of the muon and its neutrino and of charm and strange quarks. The third generation consists of the tau and its neutrino and of bottom and top quarks.

GeV. Giga electron volt, a unit of energy equal to 10^9 eV.

GLAST. Gamma-ray Large Array Satellite Telescope, a mission under consideration by NASA.

Gluon. A massless particle that carries the strong force.

Grand unified theory. Theory that unifies the electroweak force with the strong force into a single gauge theory.

Gran Sasso. Italian national underground laboratory for particle physics and astrophysics, located in a tunnel near Rome.

Granularity. Degree of fine spatial resolution in a particle detector (see Chapter 6).

Gravitational force or interaction. Weakest of the four basic forces and the one responsible for the attraction of objects to Earth and the motion of stars and planets.

Graviton. As yet undetected massless particle that carries the gravitational force.

Hadron. Subnuclear particle composed of quarks. The hadron family of particles consists of baryons and mesons. The best known are protons, neutrons, and pions.

HEPAP. Department of Energy's High-Energy Physics Advisory Panel.

HERA. Electron-proton circular collider located at DESY in Germany.

Hermeticity. Degree to which a particle detector surrounds particle interactions.

Higgs boson or *particle.* Hypothetical particle that could account for the origin of the masses of elementary particles.

High-energy physics. Another name for elementary-particle physics. This name arises from the high energies required for experiments in this field.

High-T_c superconductor. Special class of superconductors that operates at temperatures above the boiling point of liquid nitrogen, about 77 K. Ordinary superconductors operate at 20 K or lower.

Hubble constant. Measure of the expansion rate (and hence, age) of the universe, estimated from observations to be between 45 km/s per megaparsec (a megaparsec is about 3 million light-years) and 90 km/s per megaparsec.

Hypercharge. Quantum number characterizing a property of quarks and leptons.

IHEP. Institute for High-Energy Physics, a 76-GeV circular proton accelerator in Protvino, Russia.

IMB. Irvine Michigan Brookhaven proton-decay experiment.

Intermediate vector boson. General name for W and Z particles that carry the weak force.

Internal space. Mathematical space defined by possible values of particular quantum numbers.

Invariance. Property of a physical system that is unchanged when its coordinate system is altered.

Ionization. Process of removing electrons from an atom.

Isospin. Quantum number in an internal space that applies to quarks.

Isotropic. Condition of having the same property along any direction in space (i.e., rotationally invariant).

J/psi. Particle made of a *c* quark (see charm) and an anti-*c* quark. It is about three times as massive as the proton.

Jet. Narrow stream of hadrons produced in a high-energy collision.

K meson or kaon. Second least massive meson, made of one *s* quark and one *u* or *d* antiquark.

Kamiokande. Kamioka (Japan) Nucleon Decay Experiment, a 3,000-ton water-Cerenkov proton decay and neutrino detector experiment (see Super Kamiokande).

KEK. Japanese High-Energy Accelerator Research Organization's 12-GeV circular proton accelerator at Tsukuba, Japan.

Klystron. High-frequency oscillator-amplifier that uses an electron beam in a magnetic field to produce microwave radiation.

Kobayashi-Maskawa hypothesis. (See CP violation.) Hypothesis that if there are more than two generations of quark states and the states mix, then CP violation is allowed.

LAMPF. 800-MeV linear proton accelerator at Los Alamos National Laboratory, used for nuclear and elementary-particle physics.

LEP. Large Electron-Positron collider, a circular collider at CERN, Switzerland.

LEP II. LEP upgrade to an energy of 200 GeV.

Lepton. Member of the family of weakly interacting particles, which includes the electron, muon, tau, and their associated neutrinos. Leptons are acted on by the electroweak and gravitational forces, but not the strong force.

LHC. Large Hadron Collider, a particle accelerator being constructed at CERN. When completed it will be the highest-energy machine.

Lifetime. Measure of how long, on average, an unstable particle or nucleus exists before it decays.

Linac. Abbreviation for linear accelerator.

Linear accelerator. Type of accelerator in which the beam particles travel in a straight line and gain energy by "surfing" on propagating electromagnetic fields.

Lithography. In semiconductor manufacturing, using intense beams of particles or radiation to etch circuit patterns on a silicon chip.

LLNL. Lawrence Livermore National Laboratory.

Luminosity. Measure of the rate at which particles in a collider come in contact. The higher the luminosity, the greater is the rate of interactions.

MACHOs. Massive compact halo objects thought to populate the halo of the Milky Way.

MACRO. Monopole, Astrophysics, and Cosmic-ray Observatory located in the Gran Sasso laboratory.

Magnetic monopole. Hypothetical particle that would carry a single magnetic pole. All known particles with magnetic properties carry both a north and a south magnetic pole.

Main Injector. New proton accelerator used for producing antiprotons for the Tevatron as well as for a program of fixed-target physics.

Mark I. Detector at SPEAR that operated from 1972 to 1976.

Mass. Intrinsic property of a particle; the energy in a particle at rest is related to its mass by the Einstein equation $E = mc^2$. The weight of an object on Earth is proportional to its mass.

Meissner effect. Property that a superconductor will expel magnetic lines of force when cooled below its superconducting temperature.

Meson. Strongly interacting particle that is not a baryon. Mesons are composed of quark-antiquark combinations.

MeV. Mega electron volt, a unit of energy equal to 10^6 eV.

MHz. Megahertz, a frequency of 10^6 cycles per second.

Mixing. Tendency for one quantum state to transform into another.

Molecule. Type of matter made up of two or more atoms.

Monte Carlo. Statistical method that models processes caused by a series of random events.

MRI. Magnetic resonance imaging, a technique that uses the resonance of the spin of nuclei when exposed to radio waves to image the internal structure of objects.

Muon. Particle in the lepton family with a mass about 200 times that of the electron and having other properties similar to those of the electron.

Muon collider. Accelerator that collides muons with antimuons.

MWPC. Multiwire proportional chamber, a type of particle detector with many closely spaced wires, providing good spatial resolution.

NASA. National Aeronautics and Space Administration.

Neutralino. Neutral particles with a spin of one-half, predicted by supersymmetry as counterparts to the photon, the Z, and the neutral Higgs boson.

Neutrino. Electrically neutral lepton with very small, possibly zero, mass. There are at least three distinct types of neutrinos, one associated with the electron, one with the muon, and one with the tau. Neutrinos are produced in radioactive decay in stars and copiously in supernovas.

Neutrino oscillations. Neutrinos of one type may be able to change into those of another type and back again if one or more of the types have mass.

Neutron. Uncharged baryon with mass slightly greater than that of the proton. The neutron is a strongly interacting particle and a constituent of all atomic nuclei except hydrogen. An isolated neutron decays through the weak interaction to a proton, electron, and antineutrino with a lifetime of about 1,000 s.

NSF. National Science Foundation.

Nucleon. Neutron or proton.

Nucleosynthesis. Process in stars by which heavier elements are generated from lighter ones through nuclear fusion.

Nucleus. Central core of an atom, made up of neutrons and protons held together by the strong force.

Parity. Property that may be thought of in the same way as left- or right-handedness (see CP symmetry).

Particle. In high-energy physics, a component of matter on the subatomic scale.

Parton. Pointlike, internal component of hadrons, now identified as a quark or a gluon.

PEP. Positron-Electron Project, a circular collider with a maximum energy of 36 GeV, located at SLAC.

PEP-II. The successor to PEP.

PET. Positron-emission tomography. A three-dimensional imaging technique that employs characteristic gamma rays from positron annihilation.

PETRA. Electron-positron circular collider at DESY, Hamburg, Germany.

Phonon. Quantum of vibrational energy in an atomic lattice.

Photon. Quantum of electromagnetic energy. A unique massless particle that carries the electromagnetic force.

Pion. Lightest meson, consisting of a *u* or *d* quark and an anti-*u* or anti-*d* quark.

Planck scale. Scale in physics where gravity plays a role equal to other forces, at a distance, time, and energy scale of 10^{-33} cm, 10^{-43} s, and 10^{19} GeV, respectively.

Positron. Antiparticle of the electron.

Proton. Baryon with a single positive unit of electric charge and a mass approximately 1,840 times that of the electron, made of two up quarks and one down quark. The proton is the nucleus of the simple hydrogen atom and a constituent of all atomic nuclei.

PS. Circular proton accelerator with a maximum energy of 28 GeV at CERN, Switzerland.

Quantum chromodynamics (QCD). Theory that describes the strong force among quarks in a manner analogous to the description of the electromagnetic force by quantum electrodynamics.

Quantum electrodynamics (QED). Theory that describes the electromagnetic interaction in the framework of quantum mechanics. The particle carrying the electromagnetic force is the photon.

Quantum field theory. Field theory in which the fields are quantum mechanical variables.

Quantum mechanics. Mathematical framework for describing the physics at atomic and smaller length scales, where energy exists in discrete quantum units.

Quark-gluon plasma. High-energy form of nuclear matter, in which the binding forces between quarks in individual protons and neutrons decreases and creates a deconfined state characterized by the free movement of quarks and gluons throughout the nuclear volume.

Quarks. Family of elementary particles that make up hadrons. Quarks are acted on by strong, electroweak, and gravitational forces. Six are known, referred to as up, down, strange, charm, bottom, and top.

Radiation. Electromagnetic waves, energy in the form of photons.

Radiography. Technique of producing an image by using x rays or gamma rays instead of light.

Relativistic. Systems with particles moving with velocities close to the velocity of light.

RHIC. Relativistic heavy-ion collider, under construction at Brookhaven National Laboratory in New York.

Scattering. When two particles collide, they are said to scatter off each other.

Scintillation counter. Particle detector in which the passage of a charged particle produces a flash of light. This so-called scintillation light, when detected, gives the time at which the particle passed through the counter.

SLAC. Stanford Linear Accelerator Center in Stanford, California, the electron linear accelerator there having an energy of 50 GeV.

SLC. Stanford Linear Collider, at SLAC, a linear electron-positron collider with a center-of-mass energy of about 100 GeV.

SNO. Sudbury Neutrino Detector. A 1,000-ton heavy-water Cerenkov detector under construction in a mine near Sudbury, Ontario, 6,800 feet below ground. SNO is designed to detect neutrinos produced by fusion reactions in the Sun.

Soudan-II. Detector located in an underground laboratory in a mine about 1/2 mile beneath Soudan, Minnesota. Some physics goals of the experiment are to search for nucleon decay and to study atmospheric neutrino physics. The detector is a 960-ton iron calorimeter surrounded by an active shield of proportional tubes.

SPEAR. Circular electron-positron collider with center-of-mass energy of about 8 GeV that operated at SLAC in the 1970s and 1980s.

Spin. Intrinsic angular momentum possessed by a particle. Generally measured in units of the Planck constant h divided by 2π.

SPS. Super Proton Synchrotron, an accelerator at CERN.

SSC. Superconducting Super Collider.

SSRL. Stanford (California) Synchrotron Radiation Laboratory.

Standard model. Theory that summarizes the present picture of the field of elementary-particle physics. It includes three generations of quarks and leptons, the electroweak theory of weak and electromagnetic forces, and the quantum chromodynamic theory of the strong force. It does not include answers to some basic questions such as how to unify electroweak forces with the strong or gravitational forces.

Storage ring. Ring of magnets used to store circulating particles or act as a collider. Sometimes a synonym for a collider.

Strange. Third type of quark, also called s quark.

Strange particle. Particles containing at least one s quark.

String theory. Class of theories that treats elementary particles as tiny strings in a higher-dimensional space.

Superconducting magnet. See Superconductivity.

Superconducting Super Collider. Circular proton-proton collider with a center-of-mass energy of 40 TeV, located in Texas. The collider's construction was terminated by Congress in 1993.

Superconductivity. Property by which some materials, when cooled to a temperature close to absolute zero, lose all of their electrical resistance and become superconducting. Magnets with superconducting coils can produce large magnetic fields with low power costs.

Superkamiokande. Joint Japan-U.S. collaboration to construct the world's largest underground neutrino observatory. It is a water Cerenkov detector—a tank of ultrapure water 40 m in diameter and 40 m tall, viewed by thousands of phototubes—located in the Kamioka Mine, about 200 km north of Tokyo.

Superpartner. SUSY counterpart to an ordinary matter particle.

Supersymmetry. Theory of elementary particles in which each boson has a fermion counterpart and vice versa.

SUSY. See Supersymmetry.

Symmetry. General property of many objects and physical systems whereby the object or system appears unchanged when looked at from different reference frames or coordinate systems. For example, a smooth ball has spherical symmetry because it looks the same from any orientation. Many kinds of symmetry exist.

Synchrotron. Type of circular particle accelerator in which the frequency of acceleration is synchronized with the particle as it makes successive orbits.

Synchrotron radiation. Intense light or x rays emitted when electrons move in a circular orbit at relativistic speeds.

Target. Material struck by a beam of high-energy particles, used in some types of EPP experiments.

Tau. Elementary particle in the lepton family with a mass 3,500 times that of the electron but with similar properties.

Technicolor. Theory proposed to explain the masses of particles, which postulates the existence of a new strong force.

Tesla (T). Unit of magnetic field strength equal to 10,000 gauss. A modern superconducting magnet can generate a field of about 5-10 T.

TeV. Tera electron volt, a unit of energy equal to 10^{12} eV.

Tevatron. Complex of accelerator facilities at Fermilab. The main facility is a circular proton accelerator with superconducting magnets (the first large accelerator to employ such magnets) that can be used as an antiproton-proton collider with a center-of-mass energy of 2 TeV. Currently the highest-energy collider in the world.

Top. Sixth type of quark, also called t quark. The mass of the top quark is about 175 GeV.

TPC. Time projection chamber; a particle detector in which the position of the track of ionized gas left by a charged particle is detected by the time it takes for electrons in the gas to move to the ends of the chamber.

TRISTAN. Circular electron-positron collider, with center-of-mass energy of 60 to 70 GeV, that operated at the KEK laboratory in Japan in the 1980s.

Unified theories. Theories in which different forces have a common origin. For example, the electric and magnetic forces are unified in the theory of electromagnetism.

UNK. Accelerator and storage-ring complex, located at Serpukhov, Russia.

Up. One of the two lightest quarks, also called u quark. Up and down quarks form the first quark generation.

Upsilon. Meson made up of a b quark and an anti-b quark. It is approximately 10 times as massive as a proton.

VEPP-2M. Electron-positron collider near Novosibirsk, Russia.

Virtual process or particle. One that is physically forbidden in classical mechanics but allowed by quantum mechanics.

W boson. Particle that carries the charged weak force. An intermediate-vector boson with a mass of about 80 GeV.

Weak interaction. Force responsible for nuclear beta decay. Much weaker than the strong or electromagnetic forces, but stronger than gravity. The weak force has the distinctive feature that it violates parity.

WIMPs. Weakly interacting massive particles, a class of hypothetical particles thought to be a candidate for dark matter.

X rays. Photons produced when atoms in states of high-energy decay to states of lower energy.

Z boson. Particle that carries the neutral weak force. An intermediate-vector boson with a mass of about 91 GeV.

Z factory. Facility for generating large numbers of Z bosons.

DATE DUE

DATE DUE			
MAY 1 2 1998			
FEB 18 '05 S			
FEB 1 3 2005			
GAYLORD			PRINTED IN U.S.A.